3D Bioprinting

Dong-Woo Cho • Byoung Soo Kim • Jinah Jang
Ge Gao • Wonil Han • Narendra K. Singh

3D Bioprinting

Modeling In Vitro Tissues and Organs Using Tissue-Specific Bioinks

 Springer

Dong-Woo Cho
Department of Mechanical Engineering and
Division of Integrative Biosciences and
Biotechnology
Pohang University of Science and
Technology
Pohang, Korea (Republic of)

Jinah Jang
Department of Creative IT Engineering and
Mechanical Engineering
Pohang University of Science and
Technology
Pohang, Korea (Republic of)

Wonil Han
Division of Integrative Biosciences and
Biotechnology
Pohang University of Science and
Technology
Pohang, Korea (Republic of)

Byoung Soo Kim
Future IT Innovation Laboratory
Pohang University of Science and
Technology
Pohang, Korea (Republic of)

Ge Gao
Department of Mechanical Engineering
Pohang University of Science and
Technology
Pohang, Korea (Republic of)

Narendra K. Singh
Department of Mechanical Engineering
Pohang University of Science and
Technology
Pohang, Korea (Republic of)

ISBN 978-3-030-32224-3   ISBN 978-3-030-32222-9   (eBook)
https://doi.org/10.1007/978-3-030-32222-9

This Springer imprint is published by the registered company Springer Nature Switzerland AG
The registered company address is: Gewerbestrasse 11, 6330 Cham, Switzerland

Preface

In vitro tissues and organs offer a useful platform that can facilitate systematic, repetitive, quantitative investigations of drugs/cosmetics. The eventual objective of developing tissues/organs models is to reproduce structurally and physiologically relevant functions that are typically required in native human tissues/organs. The fact that animal research for cosmetic development has been inhibited for ethical reasons supports the need for more predictable, repetitive, and automated in vitro tissue/organ models with higher complexities.

3D bioprinting offers great prospects for constructing such ideal 3D tissue/organs models because it allows for reproducible and automated production of complex living tissues through spatial deposition of various cells and biomaterials. Many studies have also shown the ability of 3D bioprinting to precisely add various cells and biomaterials into a tissue construct. Along with such technical advances, the development of printable material (referred to as bioink) that is able to provide encapsulated cells with in vivo-like microenvironments is highlighted. To date, many biomaterials have been formulated as bioink, including agarose, collagen, gelatin, alginate, and fibrinogen, for this purpose. However, such homogeneous bioinks cannot completely recapitulate the complexity of native tissues/organs containing various and different kinds of proteins in respective bioprinted tissues/organs. In addition, the accumulative outcomes have demonstrated that the microenviornmental niches and complex cues within the bioink play essential roles in cellular functions and fate rather than merely holding encapsulated cells. This again emphasizes the importance of the material source being formulated as a bioink.

Decellularized extracellular matrix (dECM) has been in the spotlight as a well-qualified bionic source for closely replicating tissue-specific microenvironments, providing the diverse array of inherent ECM compositions from respective tissues/organs. Some researchers have elucidated the superiority of dECM-based bioinks through varying in vitro and in vivo evaluations. Undoubtedly, using dECM as a bioink source opens up the chance for recapitulating in vitro tissue/organ models with higher predictability. Yet, few studies have reported the prospects of 3D bioprinting in vitro tissue/organ models using dECM bioinks.

Given the above considerations, this book covers a basic understanding of in vitro tissue/organ models based on 3D bioprinting and tissue-specific dECM bioinks. We begin with the descriptions of requirements and prevalent techniques for modeling in vitro tissues/organs. After briefly explaining 3D bioprinting techniques, we move on the importance of dECM as a promising bioink source and cover its advantages and disadvantages in detail. Finally, various applications using dECM bioinks for in vitro tissue/organ modeling are outlined and discussed.

Pohang, Korea (Republic of)   Dong-Woo Cho
                                                                               Byoung Soo Kim
                                                                                     Jinah Jang
                                                                                        Ge Gao
                                                                                    Wonil Han
                                                                             Narendra K. Singh

Acknowledgements

This work was supported by the National Research Foundation of Korea (NRF) grant (NRF-2019R1A3A3005437) funded by the Korea Government and the Ministry of Education (No. 2015R1A6A3A04059015). This work was also supported by the Institute for Information & Communications Technology Promotion (IITP) grant funded by the Korea Government (MSIT) (2017-01-01982).

Contents

About the Authors

Dong-Woo Cho received his PhD in Mechanical Engineering from the University of Wisconsin-Madison in 1986. Ever since, he has been a professor in the Department of Mechanical Engineering at the Pohang University of Science and Technology. He is the director of the Center for Rapid Prototyping-based 3D Tissue/Organ Printing. His research interests include 3D microfabrication based on 3D printing technology, and its application to tissue engineering and more generally to bio-related fabrication. He has recently focused his attention on tissue/organ printing technology and development of high-performance bioinks. He serves or has served on the editorial boards of 11 international journals. He has published over 280 academic papers in various international journals in manufacturing and tissue engineering and has contributed chapters to ten books as well as writing a textbook related to tissue engineering and organ printing. He has received many prestigious awards including the "Mystery of Life" award at the Roman Catholic Archdiocese of Seoul (2018, January) and the "Respect Life" award from LINA Foundation 50+ awards (2019, April). He was also awarded the Nam-Go chair professorship at POSTECH.

Byoung Soo Kim received his PhD in Mechanical Engineering from POSTECH in 2019. During an MS and PhD integrated program at the Department of Mechanical Engineering in POSTECH under the guidance of Prof. Dong-Woo Cho, his research focused on 3D printing system development, dECM-based bioink formulation, and in vitro tissue biofabrication. He has particularly focused on 3D cell-printing of in vitro human skin models. He has suggested a novel platform for matured 3D skin constructs by employing a printable functional transwell system, rather than using commercial transwell inserts. Furthermore, he formulated a porcine skin-derived tissue-specific bioink and applied it to skin tissue engineering. A vascularized and perfusable human skin equivalent composed of an epidermis, dermis, and hypodermis was 3D cell-printed and matured. Based on his previous achievements, he is now working on in vitro skin disease modeling for better pathophysiological studies.

Jinah Jang received her PhD at the Division of Integrative Biosciences and Biotechnology at POSTECH in 2015, and she worked as postdoctoral fellow in the Department of Mechanical Engineering at POSTECH (2015–2016) and the Institute for Stem Cell and Regenerative Medicine/Department of Pathology and Bioengineering at the University of Washington (2016–2017). She joined POSTECH in the spring of 2017 as an assistant professor in creative IT engineering. Her research focuses on building the functional human tissues from stem cells via 3D bioprinting technology and printable biomaterials, particularly based on tissue-specific bioinks. Her successful achievements may lead to clinical applications for providing advanced therapeutic methods, understanding disease mechanisms, and engineering micro-tissue models.

Ge Gao received his master's degree from Huazhong University of Science and Technology (HUST) and his bachelor's degree from Harbin Institute of Technology (HIT), majoring in Materials Science. Since 2014, he has been a PhD student at the Department of Mechanical Engineering at POSTECH. His current research focuses on 3D cell-printing of vascular constructs and their applications as in vitro vascular platforms and in vivo blood vessel bypass grafts. In addition, he also developed a vascular tissue-specific bioink that can facilitate the functionalization of the fabricated vascular equivalents. These achievements might be useful for a wide range of biomedical applications, from modeling blood vessel relevant diseases to building vascularized tissues/organs.

Wonil Han received his bachelor's degree in Life Science from Handong Global University in 2015. Since then, he has been a graduate student on an MS and PhD integrated program, at the Department of Integrative Bioscience and Biotechnology at POSTECH under the supervision of Prof. Dong-Woo Cho. His research interests include liver tissue engineering and regenerative medicine via 3D cell-printing technology, the regulatory effect of dECM bioinks of normal tissues on the fibrotic disease models, and big data analysis of differential stem cell behaviors in dECM bioinks from different kinds of tissues.

Narendra K. Singh has completed his PhD in Materials Science from the Indian Institute of Technology-Banaras Hindu University (IIT-BHU), India, in 2012 and has subsequently joined as a post-doctoral research fellow in the Department of Polymer Science and Engineering, Sungkyunkwan University, South Korea (2013–2015). He is currently a postdoctoral research fellow under prof. Dong-Woo Cho at the Department of Mechanical Engineering, Pohang University of Science and Technology (POSTECH), South Korea. His research interests include biomaterials, proteomics analysis of dECM bioinks, 3D cell-printing technology, and regenerative medicine. In particular, his current research focuses on the 3D cell-printing of microfluidic kidney-on-a-chip for the development of in vitro disease models and drug toxicology advancement. He has published 12 research articles in internationally reputed peer-reviewed journals and two book chapters. His research work has been selected for the cover page of Nanosci. Nanotechnol. Lett, 1: 52–56 (2009) and his work (J. Mat. Chem. 21:15919–15927 (2011)) has been highlighted in *Nature India*.

Chapter 1
Introduction

In the past, the ability to understand the formation, function, and pathology of tissues/organs was mostly dependent on 2D cell culture [1, 2]. However, a drawback of 2D cell culture-based studies is that cells grown in 2D condition are substantially different in their morphology, and in cell–cell and cell–matrix interactions, from those grown in physiologically more relevant 3D environments [3]. This supports the fact that the 2D environment cannot be used to represent the 3D environment in understanding the situation taking place in our body. As an alternative, animal models have been used as a testing platform owing to their similarities with regard to morphology and cell–matrix interactions in 3D as well as bulky supplies. Although they offer plausible results demonstrating the importance of specific molecules and processes, there have still been discrepancies in gene ablation and chemogenomics [4]. Around 50% of drugs that pass preclinical tests turn out to be toxic for humans. As a representative example, researchers at a German pharmaceutical company firstly discovered that thalidomide could relieve morning sickness in pregnant women. After rigorous validation via animal experiments including dogs, cats, rats, hamsters, and chickens, this drug was marketed in 1957 [5]. However, thalidomide was found to cause deformity in children born to mothers who took the drug; the babies were born with missing or abnormal limbs, feet, or hands. It turned out to be toxic for humans and was withdrawn from the market. This incident is vividly remembered as a tragedy showing the inaccuracy and uncertainty of animal experiments, giving us a valuable lesson in drug development. However, this also means that some of the drugs may be nontoxic and effective for humans even if they fail in animal models [6]. This causes rejection of potentially important drugs even before they reach clinical trials. Moreover, animal models are often ineffective in reproducing features of human tumors and autoimmune diseases, which are related with physiological processes, due to the fundamental differences in the evolution of two complex systems. Furthermore, most scientific research involving the use of animals since 2013 has begun with an ethical focus, motivating researchers to find an appropriate substitute for more effective studies [7].

© Springer Nature Switzerland AG 2019   1
D.-W. Cho et al., *3D Bioprinting*, https://doi.org/10.1007/978-3-030-32222-9_1

In vitro 3D tissue models provide an excellent alternative to traditional 2D cell cultures and animal testing [8–10]. Such models are under development to provide a means of systematic, repetitive, and quantitative investigation of cell or tissue physiology in pharmaceutics and cosmetics industries. It can be more precise, usually less expensive, and less time-consuming than animal models. An increasing use of human cell-based 3D models that exactly mimic specific features of native human tissues could promote advances in understanding tissue morphogenesis and also facilitate the screening of new therapeutics. Many results have shown that in vitro models can represent and recapitulate the spatial and chemical complexity of living tissues more effectively than 2D models or animal models [11]. However, current in vitro models consisting of one to two cell types lack multiscale architectures and tissue–tissue interfaces which are of importance in creating the functional system in health and disease [12]. In particular, most of the 3D tissue models suffer from nutrition supply occurring between vascular endothelium and surrounding connective tissue and parenchymal cells, which is critical to the function and lifespan of all organs. Microfluidic organs on chips (also referred to as microphysiological systems) offer the chance to overcome such limitation by incorporating cell culture devices with continuously perfused chambers [13, 14]. Yet, they have still difficulties in creating complex multiscale architecture owing to their planar fabrication technology [12]. For this reason, such chips merely focus on analyzing the basic functions and responses of single tissues.

3D bioprinting enables the recapitulation of complex multiscale architecture by precisely positioning various cell types, biomolecules, and biomaterials simultaneously at predefined locations [15, 16]. While still in their early stages, 3D bioprinting strategies have demonstrated their potential use in development of 3D functional living in vitro tissue models [17–20]. In parallel to these technological advances, biomaterial development able to provide printed cells with favorable microenvironments is emerging as a major challenge [21, 22]. Supporting cells in our body, the native extracellular matrix (ECM) is topographically complex and contains various kinds of protein. This structure provides cell–ECM interactions and, thereby, determines important signals for cell fate [23]. To date, various biomaterials have been formulated as bioinks, including collagen, gelatin, alginate, fibrin, and silk [24]. However, due to the complex nature of native ECM, using one or more components in a bioink formulation cannot completely recapitulate the native ECM characteristics. The ideal condition would expose cells to natural microenvironments similar to the environment in which the cells are isolated. Decellularized ECM as a bioink source might be well fitted for this purpose, as long as effective separation of cellular components from native tissues is achieved while maintaining the ECM composition [25, 26]. With advances in decellularization techniques, dECM materials are currently under the spotlight as a powerful source of bioink towards advanced in vitro tissue/organ modeling [27].

In this textbook, we provide fundamental knowledge for beginners on modeling 3D in vitro tissues and organs, especially on the basis of 3D bioprinting technology. We begin with the descriptions of prerequisites (from pre-processing to post-processing) for modeling 3D tissues and organs in vitro. Regarding biofabrication of such in vitro tissue/organ models, prevalent technologies and their limitations are introduced. We next move on 3D bioprinting techniques, currently used bioinks, and their differences. Importantly, dECM materials to be utilized as bioink sources are discussed, including decellularization methods and validation. Subsequently, various applications of 3D bioprinting in vitro tissue/organ models, particularly concentrated on using dECM-based bioinks, are introduced. Finally, based on the aforementioned issues, future prospects are discussed.

References

1. Imamura Y, Mukohara T, Shimono Y, Funakoshi Y, Chayahara N, Toyoda M, et al. Comparison of 2D-and 3D-culture models as drug-testing platforms in breast cancer. Oncol Rep. 2015;33:1837–43.
2. Doke SK, Dhawale SC. Alternatives to animal testing: a review. Saudi Pharmaceut J. 2015;23:223–9.
3. Snouwaert JN, Brigman KK, Latour AM, Malouf NN, Boucher RC, Smithies O, et al. An animal model for cystic fibrosis made by gene targeting. Science. 1992;257:1083–8.
4. Dick IP, Scott RC. Pig ear skin as an in-vitro model for human skin permeability. J Pharm Pharmacol. 1992;44:640–5.
5. Vargesson N. Thalidomide-induced teratogenesis: history and mechanisms. Birth Defects Res C Embryo Today. 2015;105:140–56.
6. Emami J. In vitro-in vivo correlation: from theory to applications. J Pharm Pharm Sci. 2006;9:169–89.
7. Adler S, Basketter D, Creton S, Pelkonen O, Van Benthem J, Zuang V, et al. Alternative (non-animal) methods for cosmetics testing: current status and future prospects—2010. Arch Toxicol. 2011;85:367–485.
8. Vailhé B, Vittet D, Feige J-J. In vitro models of vasculogenesis and angiogenesis. Lab Invest. 2001;81:439.
9. Kang JH, Gimble JM, Kaplan DL. In vitro 3D model for human vascularized adipose tissue. Tissue Eng Part A. 2009;15:2227–36.
10. Baker M. Tissue models: a living system on a chip. Nature. 2011;471:661.
11. Elliott NT, Yuan F. A review of three-dimensional in vitro tissue models for drug discovery and transport studies. J Pharm Sci. 2011;100:59–74.
12. Pati F, Gantelius J, Svahn HA. 3D bioprinting of tissue/organ models. Angew Chem Int Ed. 2016;55:4650–65.
13. Bhise NS, Ribas J, Manoharan V, Zhang YS, Polini A, Massa S, et al. Organ-on-a-chip platforms for studying drug delivery systems. J Control Release. 2014;190:82–93.
14. Lee H, Cho D-W. One-step fabrication of an organ-on-a-chip with spatial heterogeneity using a 3D bioprinting technology. Lab Chip. 2016;16:2618–25.
15. Mandrycky C, Wang Z, Kim K, Kim D-H. 3D bioprinting for engineering complex tissues. Biotechnol Adv. 2016;34:422–34.
16. Murphy SV, Atala A. 3D bioprinting of tissues and organs. Nat Biotechnol. 2014;32:773–85.
17. Zhang YS, Arneri A, Bersini S, Shin S-R, Zhu K, Goli-Malekabadi Z, et al. Bioprinting 3D microfibrous scaffolds for engineering endothelialized myocardium and heart-on-a-chip. Biomaterials. 2016;110:45–59.
18. Homan KA, Kolesky DB, Skylar-Scott MA, Herrmann J, Obuobi H, Moisan A, et al. Bioprinting of 3D convoluted renal proximal tubules on perfusable chips. Sci Rep. 2016;6:34845.
19. Bhise NS, Manoharan V, Massa S, Tamayol A, Ghaderi M, Miscuglio M, et al. A liver-on-a-chip platform with bioprinted hepatic spheroids. Biofabrication. 2016;8:014101.

20. Knowlton S, Yenilmez B, Tasoglu S. Towards single-step biofabrication of organs on a chip via 3D printing. Trends Biotechnol. 2016;34:685–8.

21. Hölzl K, Lin S, Tytgat L, Van Vlierberghe S, Gu L, Ovsianikov A. Bioink properties before, during and after 3D bioprinting. Biofabrication. 2016;8:032002.

22. Kesti M, Müller M, Becher J, Schnabelrauch M, D'Este M, Eglin D, et al. A versatile bioink for three-dimensional printing of cellular scaffolds based on thermally and photo-triggered tandem gelation. Acta Biomater. 2015;11:162–72.

23. Hynes RO. The extracellular matrix: not just pretty fibrils. Science. 2009;326:1216–9.

24. Chimene D, Lennox KK, Kaunas RR, Gaharwar AK. Advanced bioinks for 3D printing: a materials science perspective. Ann Biomed Eng. 2016;44:2090–102.

25. Kim BS, Kim H, Gao G, Jang J, Cho D-W. Decellularized extracellular matrix: a step towards the next generation source for bioink manufacturing. Biofabrication. 2017;9:034104.

26. Pati F, Jang J, Ha D-H, Kim SW, Rhie J-W, Shim J-H, et al. Printing three-dimensional tissue analogues with decellularized extracellular matrix bioink. Nat Commun. 2014;5:3935.

27. Choudhury D, Tun HW, Wang T, Naing MW. Organ-derived decellularized extracellular matrix: a game changer for bioink manufacturing? Trends Biotechnol. 2018;36:787.

Chapter 2
In Vitro Modeling 3D Tissues and Organs

2.1 Prerequisites for Modeling 3D Tissues and Organs

In native tissues or organs, cells are organized in the 3D microenvironment, which allows the cell–cell and cell–extracellular matrix (ECM) interactions [1]. In this regard, 3D in vitro tissue models can better mimic the natural tissue or organ when compared to the traditional 2D in vitro cell culture platforms.

The fundamental requirement of in vitro tissue/organ models is the appropriate designs for recapitulating the natural microenvironment of native tissues/organs. A major concern of the in vitro tissue/organ models is how to replicate their functionality. Since every single tissue/organ has unique features, recapitulating respective tissues and organs is challenging. In this regard, the prerequisites for emulating 3D in vitro tissues/organs can be divided into several fundamental elements (Fig. 2.1). This chapter discusses the remodeling of in vitro 3D tissue/organ models and gives a brief overview of prerequisite components with various dimensions such as biomaterials, cell sources, biomimicry, biofabrication, vascularization, and bioreactors.

2.2 Biomimicry

The biomimicry has been applied in several fields such as civil engineering [2], aeronautical engineering [3], materials research [4], and nanotechnology [5]. Biomimicry in biofabrication (3D bioprinting) is aimed at gaining biofunctional accuracy in the fabricated tissues/organs. Such complex heterogeneous tissues have to be mimicked along with mature vasculature tree, circulatory network, and neural structures from native conditions. Furthermore, for successful replication of heterogeneous biological tissues, many noncellular factors have to be considered such as ECM composition, the gradient of soluble or insoluble micronutrients, hierarchy of

© Springer Nature Switzerland AG 2019
D.-W. Cho et al., *3D Bioprinting*, https://doi.org/10.1007/978-3-030-32222-9_2

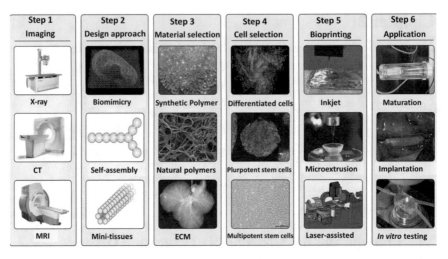

Fig. 2.1 A typical process for bioprinting 3D tissues. Imaging of the damaged tissue and its environment can be used to guide the design of bioprinted tissues. Biomimicry, tissue self-assembly, and mini-tissue building blocks are design approaches used singly and in combination. The choices of materials and cell source are essential and specific to the tissue form and function. Common materials include synthetic or natural polymers and decellularized ECM. Cell sources may be allogeneic or autologous. These components have to integrate with bioprinting systems such as inkjet, microextrusion, or laser-assisted printers. Some tissues may require a period of maturation in a bioreactor before transplantation. Alternatively, the 3D tissue may be used for in vitro applications. (Reproduced with permission from [1])

supportive cells, and 3D microenvironment (Fig. 2.1). Hence, the deep understanding of the advantages and shortcomings of these factors is necessary for recapitulating the proper functionalities of heterogeneous tissues. In addition, external environmental factors (e.g., pressure, temperature, and electrical forces), as well as biomaterials, also play a critical role in defining cell morphology and size. To control the environmental factors, the utilization of bioreactors is crucial, usually enabling the creation of similar stimuli that are essential to mimic the microenvironment of the target tissue.

2.3   Preprocessing

Reproducing the heterogeneous and complex architecture of tissues/organs requires a detailed understanding of each component's composition and their organization. First, the pre-processing phase collects the data on 3D anatomical structure and function of the target tissue or organ by using medical imaging technologies, which are usually noninvasive magnetic resonance imaging (MRI) and computed tomography (CT). Imaging data acquired from these noninvasive modalities are subsequently translated to 3D computational models using computer-aided design/manufacturing (CAD/CAM) tools (e.g., SOLIDWORKS, Autodesk123D, and CATIA) [1].

2.4 Biofabrication Process

The biofabrication process aims to mimic the complexity of tissues/organs [6]. During the biofabrication process, the cells are encapsulated into the bioinks and printed into the predefined 3D space using automated computer-controlled techniques for the development of in vitro tissue model [6]. In this perspective, several approaches have been proposed to fabricate cellular systems such as soft lithography [7], photolithography [8], microcontact printing [9], and 3D bioprinting [10, 11].

2.5 Biomaterials

The construct of in vitro tissue/organ models should be designed precisely to mimic the inherent structure of the native tissue. Using biomaterials can offer the 3D microenvironment allowing for the encapsulation of cells and relevant bioactive factors. The most relevant biomaterials for the in vitro tissue/organ fabrication must be selected carefully for guiding cellular functional behavior. ECM is the proper biomaterial to enhance cell attachment, proliferation, maturation, and ECM synthesis [12]. It can provide structural support to maintain the predefined 3D structure, while cells endure tissue reorganization during tissue development.

The mechanical integrity of engineered constructs should be similar to those of target tissue to be remodeled, in both healthy and diseased conditions. Therefore, the selected biomaterials should provide appropriate mechanical integrity during the initial development of tissues/organs. In particular, the diseased tissues show altered ECM properties, such as different cellular architecture and physicochemical properties of osteoporotic bone, compared to its normal tissue [13]. Hence, to engineer an in vitro disease model, ECM properties also should be modified from a normal case.

To date, three different categories of materials have been widely used for engineering in vitro tissues/organs. Synthetic polymers (polyglycolic acid [PGA], poly(lactic-co-glycolic acid) [PLGA], polycaprolactone [PCL]), which have tunable mechanical strength, often have been utilized in tissue engineering [14–16]. Even though the synthetic polymers have superior biocompatibility, they lack the specific binding sites to regulate the cellular signaling pathway. This weakness often hinders cell attachment and may result in cell death. Therefore, synthetic polymers have been only used as frameworks or as sacrificial inks for the bioprinting of in vitro tissue models [14–16].

Naturally derived materials usually preserve the physical and biochemical stimuli to dictate the cellular functions such as proliferation, differentiation, and maturation [12, 14]. This class of materials has been widely utilized as hydrogels containing around 90% water and supporting favorable microenvironments for encapsulated cells. Such hydrogels can be gelled through physical or chemical crosslinking processes (e.g., thermal, ionic/chemical, photo-crosslinking) [12, 14]. Their vital properties such as viscosity, gelation kinetics, stiffness, molecular weight, and concentration play important roles during fabrication of in vitro tissue models. These natural materials are further classified based on their chemical composition and functionalities such as proteins (collagen, silk, gelatin, and fibrin), glycosaminoglycans (hyaluronic acid), and polysaccharides (alginate, chitosan) [12].

Decellularized extracellular tissue matrix (dECM) is another attractive candidate to mimic the exact tissue-specific composition of the target tissue. Recently, dECMs from various tissues have been utilized as bioinks for the 3D bioprinting of in vitro tissue models [14, 17]. The fabricated constructs have shown enhanced stem cell differentiation and tissue formation in the ECM derived from the particular tissue of origin [17].

Detailed characteristics of these biomaterials will be discussed in Chaps. 5 and 6 of this textbook.

2.6 Cell Sources

To precisely recapitulate the biological features of tissue/organ in vitro, the lineage of used cells should be patient-specific or autologous [18]. Moreover, the tissues/organs contain various cell types that are responsible for specific biological functions [1]. Thus, the choice of appropriate cell type for the development of in vitro tissue/organ needs to be carefully considered [12].

Apart from the above considerations, in vitro tissue models should contain human cells to overcome the interspecies variation [20–22]. Many human primary cells are usually isolated from tissue biopsies harvested from healthy or diseased patients. However, isolation and in vitro culture of primary cells are difficult, and they have a limited life span and low proliferation potential [23]. For example, hepatocytes isolated from the human liver have a limited life span in vitro [12]. Similarly, in vitro culture and proliferation of primary glomerular podocytes are difficult, and they have limited life span in vitro, which severely hampers the in vitro modeling of functional glomerulus [24]. In this regard, stem cells have been used for the fabrication of in vitro tissues/organs due to their ability to self-renew, differentiate, and generate multiple targeted cell lineages. Nonetheless, human stem cells are utilized as the ultimate resource to fabricate the in vitro tissues/organs for drug testing and disease modeling. For example, differentiated cardiomyocytes derived from embryonic stem cells have been effectively used as a drug testing model for the screening of various types of cardioactive drugs [25]. However, there are some critical concerns of using stem cells, such as the regulation of differentiation pathways toward targeted cell lineages [19]. In addition, even though the embryonic stem cells provide a high degree of multipotency, they have some limitations in terms of difficult procurement, issues with immunogenicity, and ethical concerns [26, 27].

Induced pluripotent stem cells (iPSCs) originating from differentiated somatic cells have the characteristics of indefinite self-renewal and differentiate toward several cell phenotypes [28, 29]. Moreover, iPSCs isolated from the patient hold the hereditary features of genetic alteration; thus, iPSCs can be used for the development of in vitro disease model [23, 30]. Several other kinds of stem cells, including adult stem cells (bone marrow and adipose-derived stem cells) [31] and perinatal stem cells from amniotic fluids [32], are also utilized for tissue engineering. With established methods for their expansion and differentiation, these types of stem cells have also been utilized as cell sources for the fabrication of in vitro tissue models [18, 33, 34].

2.7 Vascularity

To biofabricate the functional tissue, the creation of the vascular network is one crucial factor [35]. In human bodies, the vascular network is needed to overcome the diffusion limit of oxygen when the thickness of tissue is above 100–200 μm [26, 36]. Without a proper vasculature network, engineered tissues would have an inherent limitation in terms of nutrient delivery, causing immature tissue formation and necrosis [37]. Thus, vasculature must be incorporated at the initial developmental stage to effectively perfuse the fabricated in vitro tissues for preventing tissue death and facilitating the formation of the endothelium. For in vitro tissue development, vascular structures must mimic the role of the native vessel, such as selective barrier function for waste and nutrients, as well as participating in homeostatic, coagulation functions [38].

2.8 Bioreactor

Suitable bioreactors are needed to provide appropriate environmental conditions for the maturation of fabricated tissues/organs models. For example, the blood vessel might require pulsatile flow; renal tubules and glomerular capillary might need the compartmentalized fluid flow. Thus, the bioreactors for blood vessel, kidney, and large tissue structures should provide fluidic flow to the structures with appropriate shear stress. The vascular bioreactor developed by Song et al. provided the fluid flow and mechanical stimulation like in vivo arteries [40]. Their study investigated that fluid flow condition showed significant changes in cell proliferation compared to the static condition. Moreover, microfluidic vascular bioreactors were also fabricated for the maturation of the printed blood vessel in a dynamic culture condition [39]. Similarly, 3D microphysiological systems were also established for the maturation of renal tubules in fluid flow condition to mimic the shear stress of the native renal tubules [41]. In the study, the fluidic condition yielded significantly enhanced tubule function compared to that of the static condition.

Other than fluidic flow, various types of signals should be provided for the maturation of tissue constructs. For example, the bioreactors for muscle and nerve tissues might need electrical stimulation; since cartilage is exposed to a continuous force, a cartilage bioreactor must provide cyclic stress [42, 43]. Takebe et al. studied the differentiation of cartilage progenitor cells to promote maturation of chondrocytes using a rotating wall vessel bioreactor [43]. A more specialized bioreactor for heart valves has also been created. This bioreactor provided chambers divided by a septum with an engineered tissue valve. This compartmentalized cylindrical-shaped bioreactor offers the appropriate pulsatile flow that regulates the proper opening and closing of the engineered valve [44].

References

1. Murphy SV, Atala A. 3D bioprinting of tissues and organs. Nat Biotechnol. 2014;32:773–85.
2. Sani MSHM, Muftah F, Siang TC. Biomimicry engineering: new area of tranformation inspired by the nature. IEEE Bus Eng Ind Appl Colloq. 2013:477–82.
3. Jakus AE, Shah RN. Multi and mixed 3D-printing of graphene-hydroxyapatite hybrid materials for complex tissue engineering. J Biomed Mater Res A. 2017;105:274–83.
4. Zhang H, Mao X, Du Z, Jiang W, Han X, Zhao D, et al. Three dimensional printed macroporous polylactic acid/hydroxyapatite composite scaffolds for promoting bone formation in a critical-size rat calvarial defect model. Sci Technol Adv Mater. 2016;17:136–48.
5. Huh D, Torisawa YS, Hamilton GA, Kim HJ, Ingber DE. Microengineered physiological biomimicry: organs-on-chips. Lab Chip. 2012;12:2156–64.
6. Moroni L, Boland T, Burdick JA, De Maria C, Derby B, Forgacs G, et al. Biofabrication: a guide to technology and terminology. Trends Biotechnol. 2018;36:384–402.
7. Khademhosseini A, Ferreira L, Blumling J III, Yeh J, Karp JM, Fukuda J, et al. Co-culture of human embryonic stem cells with murine embryonic fibroblasts on microwell-patterned substrates. Biomaterials. 2006;27:5968–77.
8. Qi H, Du Y, Wang L, Kaji H, Bae H, Khademhosseini A. Patterned differentiation of individual embryoid bodies in spatially organized 3D hybrid microgels. Adv Mater. 2010;22:5276–81.
9. Qin D, Xia Y, Whitesides GM. Soft lithography for micro- and nanoscale patterning. Nat Protoc. 2010;5:491–502.
10. Gao G, Kim BS, Jang J, Cho D-W. Recent strategies in extrusion-based three-dimensional cell printing toward organ biofabrication. ACS Biomater Sci Eng. 2019;5:1150–69.
11. Jang J, Park JY, Gao G, Cho DW. Biomaterials-based 3D cell printing for next-generation therapeutics and diagnostics. Biomaterials. 2018;156:88–106.
12. Pati F, Gantelius J, Svahn HA. 3D bioprinting of tissue/organ models. Angew Chem Int Ed Engl. 2016;55:4650–65.
13. Chen H, Zhou X, Shoumura S, Emura S, Bunai Y. Age- and gender-dependent changes in three-dimensional microstructure of cortical and trabecular bone at the human femoral neck. Osteoporos Int. 2010;21:627–36.
14. Jang J. 3D bioprinting and in vitro cardiovascular tissue modeling. Bioengineering (Basel). 2017;4:E71.
15. Kim BS, Jang J, Chae S, Gao G, Kong JS, Ahn M, et al. Three-dimensional bioprinting of cell-laden constructs with polycaprolactone protective layers for using various thermoplastic polymers. Biofabrication. 2016;8:035013.
16. Kundu J, Shim JH, Jang J, Kim SW, Cho DW. An additive manufacturing-based PCL-alginate-chondrocyte bioprinted scaffold for cartilage tissue engineering. J Tissue Eng Regen Med. 2015;9:1286–97.
17. Pati F, Jang J, Ha DH, Won Kim S, Rhie JW, Shim JH, et al. Printing three-dimensional tissue analogues with decellularized extracellular matrix bioink. Nat Commun. 2014;5:3935.

18. Leberfinger AN, Ravnic DJ, Dhawan A, Ozbolat IT. Concise review: bioprinting of stem cells for transplantable tissue fabrication. Stem Cells Transl Med. 2017;6:1940–8.

19. Caddeo S, Boffito M, Sartori S. Tissue engineering approaches in the design of healthy and pathological *in vitro* tissue models. Front Bioeng Biotechnol. 2017;5:40.

20. Griffith LG, Swartz MA. Capturing complex 3D tissue physiology *in vitro*. Nat Rev Mol Cell Biol. 2006;7:211–24.

21. Maltman DJ, Przyborski SA. Developments in three-dimensional cell culture technology aimed at improving the accuracy of *in vitro* analyses. Biochem Soc Trans. 2010;38:1072–5.

22. Pampaloni F, Stelzer EH, Masotti A. Three-dimensional tissue models for drug discovery and toxicology. Recent Pat Biotechnol. 2009;3:103–17.

23. Benam KH, Dauth S, Hassell B, Herland A, Jain A, Jang KJ, et al. Engineered *in vitro* disease models. Annu Rev Pathol. 2015;10:195–262.

24. Desrochers TM, Palma E, Kaplan DL. Tissue-engineered kidney disease models. Adv Drug Deliv Rev. 2014;69-70:67–80.

25. Harding SE, Ali NN, Brito-Martins M, Gorelik J. The human embryonic stem cell-derived cardiomyocyte as a pharmacological model. Pharmacol Ther. 2007;113:341–53.

26. Bishop ES, Mostafa S, Pakvasa M, Luu HH, Lee MJ, Wolf JM, et al. 3-D bioprinting technologies in tissue engineering and regenerative medicine: current and future trends. Genes Dis. 2017;4:185–95.

27. Yamanaka S. Induced pluripotent stem cells: past, present, and future. Cell Stem Cell. 2012;10:678–84.

28. Takahashi K, Tanabe K, Ohnuki M, Narita M, Ichisaka T, Tomoda K, et al. Induction of pluripotent stem cells from adult human fibroblasts by defined factors. Cell. 2007;131:861–72.

29. Yu J, Vodyanik MA, Smuga-Otto K, Antosiewicz-Bourget J, Frane JL, Tian S, et al. Induced pluripotent stem cell lines derived from human somatic cells. Science. 2007;318:1917–20.

30. Szebenyi K, Erdei Z, Pentek A, Sebe A, Orban TI, Sarkadi B, et al. Human pluripotent stem cells in pharmacological and toxicological screening: new perspectives for personalized medicine. Pers Med. 2011;8:347–64.

31. Pittenger MF, Mackay AM, Beck SC, Jaiswal RK, Douglas R, Mosca JD, et al. Multilineage potential of adult human mesenchymal stem cells. Science. 1999;284:143–7.

32. De Coppi P, Bartsch G Jr, Siddiqui MM, Xu T, Santos CC, Perin L, et al. Isolation of amniotic stem cell lines with potential for therapy. Nat Biotechnol. 2007;25:100–6.

33. Roseti L, Cavallo C, Desando G, Parisi V, Petretta M, Bartolotti I, et al. Three-dimensional bioprinting of cartilage by the use of stem cells: a strategy to improve regeneration. Materials. 2018;11:1749.

34. Skardal A, Mack D, Kapetanovic E, Atala A, Jackson JD, Yoo J, et al. Bioprinted amniotic fluid-derived stem cells accelerate healing of large skin wounds. Stem Cells Transl Med. 2012;1:792–802.

35. Datta P, Ayan B, Ozbolat IT. Bioprinting for vascular and vascularized tissue biofabrication. Acta Biomater. 2017;51:1–20.

36. Carmeliet P, Jain RK. Angiogenesis in cancer and other diseases. Nature. 2000;407:249–57.

37. Malda J, Woodfield TB, van der Vloodt F, Kooy FK, Martens DE, Tramper J, et al. The effect of PEGT/PBT scaffold architecture on oxygen gradients in tissue engineered cartilaginous constructs. Biomaterials. 2004;25:5773–80.

38. Michiels C. Endothelial cell functions. J Cell Physiol. 2003;196:430–43.

39. Kolesky DB, Truby RL, Gladman AS, Busbee TA, Homan KA, Lewis JA. 3D bioprinting of vascularized, heterogeneous cell-laden tissue constructs. Adv Mater. 2014;26:3124–30.

40. Song L, Zhou Q, Duan P, Guo P, Li D, Xu Y, et al. Successful development of small diameter tissue-engineering vascular vessels by our novel integrally designed pulsatile perfusion-based bioreactor. PLoS One. 2012;7:e42569.

41. Weber EJ, Chapron A, Chapron BD, Voellinger JL, Lidberg KA, Yeung CK, et al. Development of a microphysiological model of human kidney proximal tubule function. Kidney Int. 2016;90:627–37.

42. Griffin DJ, Vicari J, Buckley MR, Silverberg JL, Cohen I, Bonassar LJ. Effects of enzymatic treatments on the depth-dependent viscoelastic shear properties of articular cartilage. J Orthop Res. 2014;32:1652–7.
43. Takebe T, Kobayashi S, Kan H, Suzuki H, Yabuki Y, Mizuno M, et al. Human elastic cartilage engineering from cartilage progenitor cells using rotating wall vessel bioreactor. Transplant Proc. 2012;44:1158–61.
44. Konig F, Hollweck T, Pfeifer S, Reichart B, Wintermantel E, Hagl C, et al. A pulsatile bioreactor for conditioning of tissue-engineered cardiovascular constructs under endoscopic visualization. J Funct Biomater. 2012;3:480–96.

Chapter 3
Prevalent Technologies for In Vitro Tissue/Organ Modeling

3.1 Introduction

The emergence of novel techniques has led to the advent of 3D in vitro platforms for representing the biological and molecular processes involved in health and disease. These in vitro platforms can be used to interpret relevant pathophysiology, offer prognosis of disease progress, predict drug efficacy, and establish therapeutic strategies. In recent years, a plethora of technologies have been successfully applied to model the functions and abnormalities of various tissues and organs in vitro. This chapter focuses on the introduction of prevalent technology for 3D in vitro modeling tissues and organs.

3.2 Transwell System

The permeable transwell facilitates the co-culture of multiple types of cells to study the cellular interplays. This system is composed of a standard culture well and an insert bottomed with a porous membrane (Fig. 3.1a). This device is commercially available in a wide range of diameters, types of membrane, and pore sizes to satisfy a variety of research purposes. When the insert was placed in the well, a lower and an upper chamber separated by the cell-permeable membrane were built. By configuring the distribution of cells, biomolecules, and physical stimulations in different partitions, it helps to spatially locate specific cells to study cellular functions and interactions. For instance, in one study, two approaches of co-culture, mixed and transwell, were used to examine whether the physical cell–cell interactions or soluble secreted factors mediate the effects of trastuzumab, a drug for the treatment of breast cancer (Fig. 3.1b) [1]. The results indicate that the therapeutic efficacy of trastuzumab is not activated unless the cancer cells and immune cells (peripheral blood mononuclear cells [PBMCs]) physically interact.

© Springer Nature Switzerland AG 2019
D.-W. Cho et al., *3D Bioprinting*, https://doi.org/10.1007/978-3-030-32222-9_3

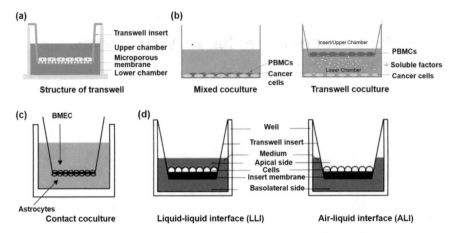

Fig. 3.1 Transwell system and its representative applications. (**a**) General structure of a transwell system. (**b**) Comparison of mixed co-culture and transwell co-culture of cancer cells and peripheral blood mononuclear cells (PBMCs). (Reproduced with permission from [1]). (**c**) Contact co-culture of brain microvascular endothelial cells (BMECs) and astrocytes. (Reproduced with permission from [2]). (**d**) Schematic diagram of liquid–liquid interface (LLI) and air–liquid interface (ALI) created using transwell systems. (Reproduced with permission from [3])

In addition to studying the cell–cell interactions, the transwell system can be used to mimic the barrier tissues. The blood-brain barrier (BBB), for example, has been modeled in vitro using a transwell system by seeding brain microvascular endothelial cells (BMECs) and astrocytes on both sides of the membrane with 0.4 μm pores (Fig. 3.1c) [2]. This contact co-culture setting-up enabled transepithelial electrical resistance measurements, leakage tests, and assays for specific BBB enzymes. Due to its similar permeability to that of the BBB in vivo, it could be a useful tool in the exploration of neuroprotective and anticancer drugs.

Moreover, using the transwell system, it is possible to emulate the biological and physical interfaces of native tissues. Owing to the chamber compartmentalization, air–liquid interface (ALI) and liquid–liquid interface (LLI) can be customized to mimic environmental stimulations for guiding cell differentiation (Fig. 3.1d) [3]. For example, the ALI has been used for modeling respiratory (lung and airway epithelial) and skin (epidermis) tissues [4, 5]. This design permits basal stem cells to grow with their basal surfaces in contact with media supplied from the lower chamber, and the top of the cellular layer is exposed to air. As a result, pseudostratified striations of respiratory epithelial and differentiated epidermis tissues can be achieved.

Despite these advantages over traditional 2D cell culture, the transwell system is confined to in vitro modeling of layered tissues because it is difficult to mirror the tissues/organs with intricate structures and heterogeneous cell distributions. In addition, this approach is usually limited in static culture, which precludes the supply of biochemical and physical stimulations (e.g., the gradient of chemical agents and dynamic flows) that are important signals tightly associated with tissue functions and diseases.

3.3   Cell Spheroids and Cell Sheets

The cell spheroids and cell sheets are formed by clusters of cells and, more importantly, the cell-secreted extracellular matrices (ECMs). These cell aggregates, compared to 2D cell culture, can resemble cell organization and interactions and thus are morphologically and functionally similar to in vivo environments.

The techniques for generating cell spheroids usually apply the hanging drop approach, which suspends cells in drops of medium hanging upside down from a surface (Fig. 3.2a), or use an adhesion-resistant cell culture surface combined with continuous agitation (Fig. 3.2b) [6]. As a result, rather than attaching on the

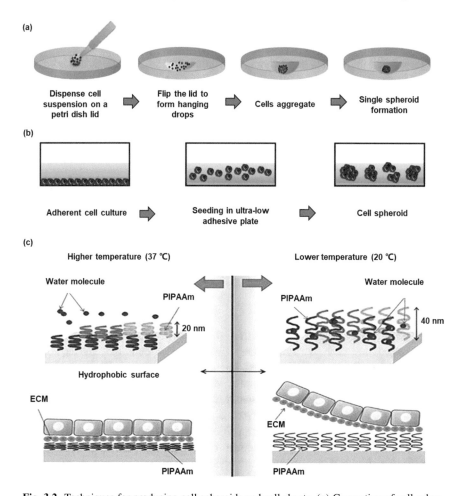

Fig. 3.2 Techniques for producing cell spheroids and cell sheets. (**a**) Generation of cell spheroids using hanging drop approach. (**b**) Cell spheroids formation using an ultralow adhesive plate. (Reproduced with permission from [6]). (**c**) Production of cell sheets using a substrate coated by thermal-responsive Poly(N-isopropylacrylamide) (PIPAAm). (Reproduced with permission from [8])

adhesive substrate, suspended cells bond to each other and synthesize ECM to form spheroids for research. The spherical shape of the cell aggregates allows investigation of the biological variations between their cores and surfaces as well as evaluating its resultant influences on microenvironments. As one representative case, tumor cell spheroids have been successfully grown in vitro, exhibiting hypoxic cores and releasing angiogenic factors. This progress facilitated their co-culture with endothelial cells to study tumor-induced angiogenesis and examine drug toxicity [7]. In recent years, the micro molding system for culturing the cell spheroids, such as AggreWell, has attracted extensive interest due to their capacity for producing spheroids with controllable and uniform dimensions. A main drawback of the technique is the difficulty in controlling the uniformity of obtained spheroids, which limits the application of standardized pharmacological tests.

Cell sheets, another promising technique for generating 3D cellularized constructs, can offer enhanced mechanical strength and improved tissue generation. Single-layered cell sheets can be gained by culturing cells in a dish coated using thermal-responsive polymer (Fig. 3.2b) [8]. Once the cells reach confluence and secrete ECMs to generate a firm layer, it is possible to harvest them as an intact cell sheet by simply reducing culture temperature to trigger the hydrophobicity changes of the polymer-coated substrates (e.g., poly(N-isopropylacrylamide) [PIPAAm]). Moreover, by stacking multiple-layered cells sheets, the well-preserved cell–cell junction, cell-surface protein, and synthesized ECMs can lead to the sheets' fusion, which results in 3D tissue models. This technique has been used to tissue-engineer layered tissues (e.g., bone, cornea, cardiac, skin, etc.) and thick vascularized tissues [9–11]. Although the cell sheets technique has been successfully used for both in vitro tissue modeling and in vivo tissue regeneration, the high cost and longtime preparation required for sheet production hinder its dissemination.

3.4 Organoids

Although similar to the cell spheroids that are in vitro generated 3D cell aggregates, the organoids are derived from primary tissue or stem cells that are capable of renewal and self-organization and exhibit organ functionality. Compared to cell spheroids, organoids are highly complex and more mimetic to in vivo circumstances. First of all, the involved self-renewal stem cells can differentiate into cells of all major cell linages with similar frequency to that in physiological condition, which provides the organoids with compositions and architectures similar to primary tissues. Moreover, due to the advancement of biology and gene engineering, it is possible to manipulate niche components and gene sequences to build the organoids that are biologically relevant to target-tissue models. In addition, since organoids can be cryopreserved, it has become a stable system for extended cultivation by leveraging self-renewal and differentiation ability of stem cells.

Based on either the isolation of primary tissues or gaining pluripotent stem cells (e.g., induced pluripotent stem cells [iPSCs] or embryonic stem cells [ESCs]),

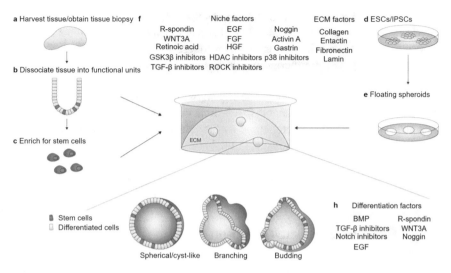

Fig. 3.3 Organoid generation and culture from primary tissue and ESCs/iPSCs. (**a**) Organoids can be generated from primary tissue that is (**b**) dissociated into functional sub-tissue units containing stem cells. (**c**) These are further digested into single cells and FACS sorted to enrich for stem cells. Both the functional units and single stem cells can give rise to organoids under the appropriate culture conditions. (**d**) Organoids derived from ESCs and iPSCs undergo directed differentiation toward the desired germ lineage, (**e**) eventually generating floating spheroids that are subsequently embedded in ECM to initiate organoid culture. (**f**) Organoids are typically cultured in an extracellular matrix (ECM) surrounded by culture media supplemented with niche factors specific to the organoid type. (**g**) Common niche and ECM factors are listed, including factors constituting the R-spondin method. (**h**) Stem cells are maintained and perpetuated in organoids, continually giving rise to differentiated progeny. Typical morphologies are classified as spherical, branching, or budding. Organoids can either differentiate spontaneously or be induced to differentiate toward desired lineages or cell types by adding suitable differentiation factors (left) and/or withdrawing factors that promote stemness (right). (Reproduced with permission from [12])

organoids can be generated by providing appropriate physical and biochemical cues (Fig. 3.3). While the physical cues, mainly ECM proteins (e.g., collagen, fibronectin, entactin, and laminin), contribute to supporting cell adhesion and survival, the biochemical cues such as growth factors and cytokines modulate signaling pathways to influence cell proliferation, differentiation, and self-renewal [12].

Since organoids are physiologically relevant and amenable to biological/molecular analyses, they hold great promise in both basic study and translational researches. The organoids originated from iPSCs and ESCs preserve cellular stemness and, thus, are useful for studying the progress of embryonic development, linage specification, and tissue homeostasis. Numerous studies have developed organoids of the brain, pancreas, and stomach to study sequential differentiation steps induced by related signaling pathways [13–16]. Moreover, the organoids include nearly all components of the organ, which helps in understanding the pathogenesis of infectious diseases as well as in investigating drug toxicity and efficacy. For instance, lung organoids derived from iPSCs from healthy child carrying null alleles of the interferon regulatory factor-7 gene have been employed to study influenza virus replication [17]. Human kidney organoids were used to demonstrate the nephrotoxicity of cisplatin [18]. Furthermore, due to the gene specification, the organoids derived from adult stem cells of individual patients may allow for the development of personalized medicine. Although the organoids are regarded as near-physiological model systems, they are still unable to mimic in vivo growth factor signaling gradients and face difficulties in emulating biomechanical stimulations that stem cells encounter in vivo. In addition, because the knowledge of the tissues whose niches factors are not well understood (e.g., ovary), it is challenging to culture the organoids reflecting these tissues and organs.

3.5   Microfluidic Tissue/Organ-on-a-Chip

As a novel analytical tool, tissue/organ-on-a-chip technology has attracted increasing interest and attention since Donald E. Ingber introduced the concept in 2010 [19]. These chips usually involve multiple microfluidic channels built in a small chip body, about the size of an AA battery, constructed from silicone, plastic, and glass. In such devices, the fluidic channels help to provide biomimetic blood/airflow stimulation, and their flexibility allows the chips to recreate breathing motions or undergo muscle contractions, which are difficult to be achieved in the models based on cell culture only. Therefore, the tissue/organ-on-a-chip can more accurately recapitulate the natural physiology and mechanical forces that cells experience in the human body. So far, various chip designs allow for the simulation of practically every human tissue/organ, including the heart, lung, kidney, bone, skin, gut, and many others (Fig. 3.4) [20].

One of the representative pioneering systems realized the reproduction of the complex physiological functionality of pulmonary epithelial cells using a lung-on-a-chip device (Fig. 3.5) [19]. This chip includes two apposed microchannels separated by a thin and flexible porous membrane where the epithelial cells and endothelial cells were co-cultured. Using a vacuum implementation, cells were mechanically simulated by the cyclic strain of breathing motions. When compared to a static version of this device or a transwell culture system, this dynamic culture significantly increases the rate of nanoparticle translocation across the porous membrane. In addition, the researchers validated the biological accuracy of the device by demonstrating the cell responses toward pulmonary inflammation and infection. In recent progress, this system has been used for modeling respiratory diseases such as chronic obstructive pulmonary diseases or lung cancer [21].

Besides the advantage of precisely controlling physiological environments, this technique holds the promise of creating a human-like system-on-a-chip by integrating multiple devices based on the success in constructing each type of human organ. Meanwhile, with the advancement of the analytical instruments and software, this tiny chip might be used in portable devices to conveniently trace the patient's health conditions, probably, using smartphone apps.

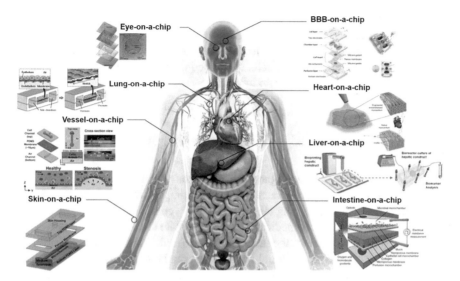

Fig. 3.4 Various organ-on-a-chip models. These models have been reported to mimic certain physiological conditions in human organs or tissues, including eye-on-a-chip, lung-on-a-chip, vessel-on-a-chip, skin-on-a-chip, BBB-on-a-chip, heart-on-a-chip, liver-on-a-chip, and intestine-on-a-chip. (Reproduced with permission from [20])

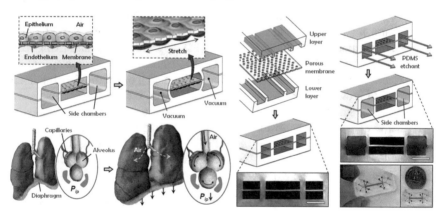

Fig. 3.5 Biologically inspired design of a human breathing lung-on-a-chip device. The microfabricated lung mimic device uses compartmentalized PDMS microchannels to form an alveolar–capillary barrier on a thin, porous, flexible PDMS membrane coated with ECM. The device recreates physiological breathing movements by applying vacuum to the side chambers and causing mechanical stretching of the PDMS membrane forming the alveolar–capillary barrier. (Reproduced with permission from [19])

On the other hand, despite the burgeoning development of tissue/organ-on-a-chip, there are a number of technical challenges that must be overcome to advance this technology. For instance, the manufacturing techniques that can produce such tiny systems should be evolved. The common fabrication issues, such as the formation of bubbles in microstructures, can be detrimental to cell survival and functions.

In addition, to introduce living cells into these devices, seeding consistency is a big concern. Different seeding results can cause varied cellular activity and analytical outcomes. Moreover, the materials for creating the chips need to be deliberated. It is usually difficult to ensure the cells will function regularly within a synthesized matrix in the same way that they do in vivo.

3.6   3D Bioprinting of In Vitro Tissue/Organ Models

The tissue-engineered tissue/organ equivalents have emerged as promising 3D in vitro models that can closely recapitulate the pathophysiology of relevant targets. Conventional tissue engineering techniques usually combine scaffolds and cells to form new viable tissues for medical purposes. Despite significant achievements in constructing artificial tissues, the inability of these traditional methods to construct structures with precise architectural features and spatial location of cells has motivated the search for new alternatives. One promising technique for the fabrication of 3D tissue/organ analogs is 3D bioprinting.

As an additive manufacturing method, the 3D bioprinting uses special bioprinters to deposit multiple cell-laden biomaterials layer-by-layer directed by digital information from 3D computer-aided design (CAD) files. As a result, it can produce complex structures with varied compositions that are biomimetic to their natural counterparts for the analytical use and drug toxicity prediction (Fig. 3.6) [22]. Unlike the traditional 3D printing techniques, 3D bioprinting integrates living cells, growth factors, and biomaterials to biomedical parts that maximally imitate natural tissue characteristics. To sustain the survival and function of cells and maintain the bioactivity of growth factors, the printing process needs to be biologically amiable. Until now, several types of 3D bioprinting technique have been developed to construct biologically complex tissue analogs, including inkjet printing, 3D extrusion, and laser-assisted 3D printing methods [23].

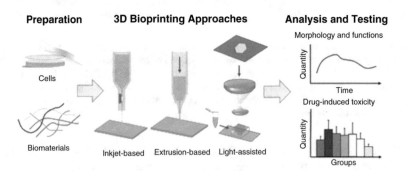

Fig. 3.6 Schematic diagram showing the 3D bioprinting technique and relevant applications of produced tissue-mimicking constructs. (Reproduced with permission from [22])

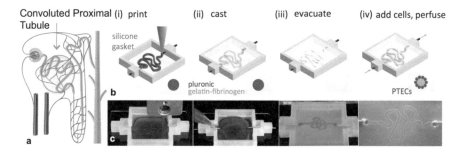

Fig. 3.7 3D-convoluted renal proximal tubule-on-a-chip. (**a**) Schematic of a nephron highlighting the convoluted proximal tubule, (**b**, **c**) corresponding schematics and images of different steps in the fabrication of 3D-convoluted, perfusable proximal tubules, in which a fugitive ink is first printed on a gelatin-fibrinogen extracellular matrix (ECM) (i), additional ECM is cast around the printed feature (ii), the fugitive ink is evacuated to create an open tubule (iii), and PTEC cells are seeded within the tubule and perfused for longtime periods (iv). (Reproduced with permission from [25])

Using these techniques, pioneering studies have successfully produced a variety of in vitro models that accurately replicate the physiological functions of natural tissues, including liver, heart, muscle, vascularized tissue, and cancer models [24]. For example, one recent study fabricated a 3D model of renal proximal tubule [25]. By removing the 3D-bioprinted fugitive materials in a casted ECM hydrogel, a perfusable-convoluted microchannel built-in matrix was successfully achieved (Fig. 3.7). Upon seeding of proximal tubule epithelial cells (PTECs) and implementation of perfusion signals, the engineered model exhibited significantly enhanced epithelial morphology and functional properties, which are difficult to observe in 2D cell culture. After introducing nephrotoxin, cyclosporine A, the epithelial barrier is disrupted in a dose-dependent manner.

3D bioprinting technology presents the capability of precisely positioning biomaterials to reconstruct complex structures that can be utilized for disease modeling and tissue function understanding. However, challenges still remain to fully recapitulate the cellular organization and structural mimicry of native tissue/organs. The insufficient resolution is the first limitation of 3D bioprinting. Currently, only laser-assisted approaches, such as stereolithography, can reach microscale resolution [26]. Considering the sophisticated and ultrafine architecture of human tissues (e.g., the capillaries in the human body are within microscale) [27], it is necessary to improve the resolution of 3D bioprinting techniques. In addition, the fabrication efficiency is limited in all of the current 3D bioprinting techniques. It has been reported that a bioprinter needs to operate for several hours to fabricate scalable organ equivalents, such as a mouse liver [28]. Considering the potential damage to cells during long-term fabrication, it is necessary to minimize the printing time for producing in vitro tissue/organ models with full biological functions. More details of 3D bioprinting techniques are described in Chap. 4 of this book.

References

1. Shi Y, Fan X, Meng W, Deng H, Zhang N, An Z. Engagement of immune effector cells by trastuzumab induces HER2/ERBB2 downregulation in cancer cells through STAT1 activation. Breast Cancer Res. 2014;16(2):R33.
2. Wang Y, Wang N, Cai B, Wang G-y, Li J, Piao X-x. In vitro model of the blood-brain barrier established by co-culture of primary cerebral microvascular endothelial and astrocyte cells. Neural Regen Res. 2015;10(12):2011.
3. Lee Y, Dizzell S, Leung V, Nazli A, Zahoor M, Fichorova R, Kaushic C. Effects of female sex hormones on susceptibility to HSV-2 in vaginal cells grown in air-liquid interface. Viruses. 2016;8(9):241.
4. Lin H, Li H, Cho HJ, Bian S, Roh HJ, Lee MK, Kim JS, Chung SJ, Shim CK, Kim DD. Air-liquid interface (ALI) culture of human bronchial epithelial cell monolayers as an in vitro model for airway drug transport studies. J Pharm Sci. 2007;96(2):341–50.
5. Li L, Fukunaga-Kalabis M, Herlyn M. The three-dimensional human skin reconstruct model: a tool to study normal skin and melanoma progression. J Vis Exp. 2011;(54):e2937.
6. Hoarau-Véchot J, Rafii A, Touboul C, Pasquier J. Halfway between 2D and animal models: are 3D cultures the ideal tool to study cancer-microenvironment interactions? Int J Mol Sci. 2018;19(1):181.
7. Kelm JM, Sanchez-Bustamante CD, Ehler E, Hoerstrup SP, Djonov V, Ittner L, Fussenegger M. VEGF profiling and angiogenesis in human microtissues. J Biotechnol. 2005;118(2):213–29.
8. Li M, Ma J, Gao Y, Yang L. Cell sheet technology: a promising strategy in regenerative medicine. Cytotherapy. 2019;21:3.
9. Matsuda N, Shimizu T, Yamato M, Okano T. Tissue engineering based on cell sheet technology. Adv Mater. 2007;19(20):3089–99.
10. Sasagawa T, Shimizu T, Sekiya S, Haraguchi Y, Yamato M, Sawa Y, Okano T. Design of pre-vascularized three-dimensional cell-dense tissues using a cell sheet stacking manipulation technology. Biomaterials. 2010;31(7):1646–54.
11. Haraguchi Y, Shimizu T, Yamato M, Okano T. Scaffold-free tissue engineering using cell sheet technology. RSC Adv. 2012;2(6):2184–90.
12. Fatehullah A, Tan SH, Barker N. Organoids as an in vitro model of human development and disease. Nat Cell Biol. 2016;18(3):246.
13. Clevers H. Modeling development and disease with organoids. Cell. 2016;165(7):1586–97.
14. Yin X, Mead BE, Safaee H, Langer R, Karp JM, Levy O. Engineering stem cell organoids. Cell Stem Cell. 2016;18(1):25–38.
15. Lancaster MA, Knoblich JA. Generation of cerebral organoids from human pluripotent stem cells. Nat Protoc. 2014;9(10):2329.
16. Drost J, Clevers H. Organoids in cancer research. Nat Rev Cancer. 2018;18(7):407.
17. Ciancanelli MJ, Huang SX, Luthra P, Garner H, Itan Y, Volpi S, Lafaille FG, Trouillet C, Schmolke M, Albrecht RA. Life-threatening influenza and impaired interferon amplification in human IRF7 deficiency. Science. 2015;348(6233):448–53.
18. Takasato M, Pei XE, Chiu HS, Maier B, Baillie GJ, Ferguson C, Parton RG, Wolvetang EJ, Roost MS, de Sousa Lopes SMC. Kidney organoids from human iPS cells contain multiple lineages and model human nephrogenesis. Nature. 2015;526(7574):564.
19. Huh D, Matthews BD, Mammoto A, Montoya-Zavala M, Hsin HY, Ingber DE. Reconstituting organ-level lung functions on a chip. Science. 2010;328(5986):1662–8.
20. Li X, Tian T. Recent advances in an organ-on-a-chip: biomarker analysis and applications. Anal Methods. 2018;10(26):3122–30.
21. Konar D, Devarasetty M, Yildiz DV, Atala A, Murphy SV. Lung-on-a-chip technologies for disease modeling and drug development: supplementary issue: image and video acquisition and processing for clinical applications. Biomed Eng Comput Biol. 2016;7:BECB.S34252.

22. Ma X, Liu J, Zhu W, Tang M, Lawrence N, Yu C, Gou M, Chen S. 3D bioprinting of functional tissue models for personalized drug screening and in vitro disease modeling. Adv Drug Deliv Rev. 2018;132:235–51.
23. Murphy SV, Atala A. 3D bioprinting of tissues and organs. Nat Biotechnol. 2014;32(8):773.
24. Mandrycky C, Wang Z, Kim K, Kim D-H. 3D bioprinting for engineering complex tissues. Biotechnol Adv. 2016;34(4):422–34.
25. Homan KA, Kolesky DB, Skylar-Scott MA, Herrmann J, Obuobi H, Moisan A, Lewis JA. Bioprinting of 3D convoluted renal proximal tubules on perfusable chips. Sci Rep. 2016;6:34845.
26. Gao G, Soo Kim B, Jang J, Cho DW. Recent strategies in extrusion-based three-dimensional cell printing toward organ biofabrication. ACS Biomater Sci Eng. 2019;5:1150.
27. Datta P, Ayan B, Ozbolat IT. Bioprinting for vascular and vascularized tissue biofabrication. Acta Biomater. 2017;51:1–20.
28. Marcos R, Monteiro RA, Rocha E. Design-based stereological estimation of hepatocyte number, by combining the smooth optical fractionator and immunocytochemistry with anti-carcinoembryonic antigen polyclonal antibodies. Liver Int. 2006;26(1):116–24.

Chapter 4
3D Bioprinting Techniques

4.1 Extrusion-Based 3D Cell Printing

Extrusion printing uses either an air-force pump or a mechanical screw plunger to dispense bioinks (Fig. 4.1a). The extrusion applies a continuous force, enabling printing of uninterrupted cylindrical lines rather than a single bioink droplet. The biggest strength is that most types of hydrogel prepolymer solutions with a wide range of viscosities (30 mPa/s to $> 6 \times 10_7$ mPa/s) [2] as well as cell aggregates with high cell density can be printed out with the extrusion printers. For this reason, nearly 30,000 3D printers are on the market worldwide, and academic institutions are also applying microextrusion-based printing for tissue and organ modeling research [3]. While extrusion is under the most spotlight due to these advantages, it exposes the encapsulated cells to significant mechanical stresses that might reduce cell viability; reportedly, cell viability (40–86%) after extrusion printing is lowest compared with those of other printing techniques. Furthermore, printer resolution and speed are basically lower than inkjet printings. Although cell viability can be increased using lower pressures and large nozzle sizes, the trade-off may be a major loss of resolution and print speed.

Most existing commercial 3D cell printers, including the Bioplotter (EnvisionTec, Gladbeck, Germany) and NovoGen 3D Bioprinting platform (Organovo, San Diego, USA), are based on extrusion technology. Extrusion printing has provided good compatibility with photo, chemical, as well as thermal crosslinkable hydrogels of very different viscosities at a reasonable cost [4]. A typical extrusion cell printing system, MtoBs, developed by our group includes three-axis motion control with six dispensing heads, supporting up to six different bioinks [5]. A further study upgraded MtoBS to include heating and cooling functions to control thermally sensitive hydrogel at substrate plate and printing heads [6]. These achievements facilitated the fabrication of large volumetric 3D cell-printed constructs with high cell viability.

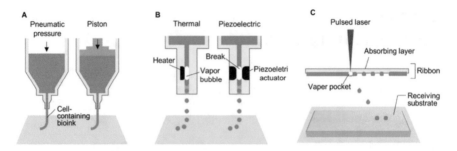

Fig. 4.1 Schematic illustration of (**a**) extrusion, (**b**) inkjet, (**c**) laser-assisted printing with different working principles. (Reprinted with permission from [1]. Copyright (2016) American Chemical Society)

4.2   Inkjet-Based 3D Cell Printing

The inkjet-based printing technique was firstly applied for tissue engineering and is very similar to conventional 2D inkjet printing [7]. A hydrogel prepolymer solution encapsulating cells (i.e., a bioink) is loaded in the ink cartridge. The cartridge is then connected to a printing head. During the printing, the printing head is exerted by a thermal or piezoelectric actuator, which could print the bioink out while generating droplets of a controllable size (Fig. 4.1b). The advantages of inkjet printing include low cost, high printing speed and resolution, and relatively high cell viability (usually 80–90%) [8]. However, because current printing heads are based on microelectromechanical systems (MEMs) devices, a relatively small deformation is generated by either thermal or piezoelectric actuation at the nozzle opening. Hence, these printing heads cannot squeeze out high-viscosity material (>15 mPa/s). Moreover, such small deformation does not support high cell density (>1 × 10^6 cells/mL), since high cell density increases the average viscosity of bioinks and eventually leads to clogging of the nozzle. Hence, the biological materials have to be in a liquid form to allow droplet formation with low viscosity. Given this feature, one common drawback is the difficulty in stacking up 3D solid constructs without immediate crosslinking processes, meaning that rapid crosslinking processes, such as chemical or ultraviolet mechanisms, are required soon after deposition. For realizing this, biomaterials need to be chemically modified, leading to decreased viability and functionality. Recent work has been focused on another disadvantage of inkjet printing, referred to as the settling effect [7]. Although bioinks loaded in the ink cartridge are well mixed with cells in the first place, the cells begin to settle on the lower position of the cartridge during the printing. It again increases the viscosity of the bioink at the nozzle opening, resulting in clogging of the printing nozzle.

4.3 Laser-Assisted 3D Cell Printing

Laser-assisted printing is originated from laser direct-write and laser-induced transfer technologies [9, 10]. The critical part of the laser-assisted printing system is a donor layer that responds to laser stimulation. The donor layer comprises a "ribbon" structure containing an energy-absorbing layer on the top and a layer of bioink solution suspended on the bottom. During the printing, a focused laser pulse is used to stimulate a small area of the absorbing layer. This laser pulse vaporizes a portion of the donor layer, creating a high-pressure bubble at the interface of the bioink layer and propelling the suspended bioink. The dropping bioink droplet is collected on the receiving substrate and subsequently crosslinked. Compared with inkjet printing, laser-assisted printing can avoid direct contact between the dispenser and the bioinks. This noncontact printing technique does not provoke mechanical stress in the cells, which results in relatively high cell viability (>95%). Additionally, laser-assisted printing can print high-viscosity materials, which means that laser-based printing can use more types of bioinks than inkjet printing. While these features are promising for engineering complex in vitro tissues and organs, the side effects of laser exposure on the cells are not yet fully understood. Furthermore, the parts used for printing including laser diodes with high resolution and intensity are considerably expensive when compared to other bioprinting techniques. Moreover, the fact that manipulation of the laser printing system is complex and cumbersome makes it difficult for users to adapt the technique for tissue engineering research. For this reason, laser-assisted printing is somewhat less common when compared with other printing techniques.

4.4 Conclusion

Obviously, 3D bioprinting opens up the chance for engineering complex tissues and organs in vitro by enabling the precise deposition of various cells and biomaterials at predefined positions. Working principles of each printing module are introduced, and different features are highlighted in light of cell viability, types of applicable bioinks, and resolution. The comparison of bioprinting technologies is summarized in Table 4.1. Considering these important different features, researchers need to scrutinize respective techniques and select one of the three techniques for ideal 3D bioprinting and modeling target tissues and organs.

Table 4.1 Comparison of 3D bioprinting techniques along with working principles

Types of 3D bioprinting system				
	Inkjet	Extrusion	Laser-assisted	Refs.
Bioink viscosities	3.5–12 mPa/s	30 mPa/s to >6 × 10^7 mPa/s	1–300 mPa/s	[11–13]
Gelation	Chemical, photo-crosslinking	Chemical, photo-crosslinking, sheer thinning, temperature	Chemical, photo-crosslinking	[10, 14]
Preparation time	Low	Low to medium	Medium to high	[15]
Printing speed	Fast (1–10,000 droplets per second)	Slow (10–50 μm/s)	Medium (1600–2000 mm/s)	[16]
Cell viability	<1 pl to >300 pl droplets, 50 μm wide	5 μm to millimeter wide	Microscale resolution	[17, 18]
Cell densities	>85%	40–80%	>95%	[19, 20]
Cost	Low	Medium	High	[21]

References

1. Jang J, Yi H-G, Cho D-W. 3D printed tissue models: present and future. ACS Biomater Sci Eng. 2016;2:1722–31.
2. Gao G, Kim BS, Jang J, Cho D-W. Recent strategies in extrusion-based three-dimensional cell printing toward organ biofabrication. ACS Biomater Sci Eng. 2019;5:1150–69.
3. Ozbolat IT, Hospodiuk M. Current advances and future perspectives in extrusion-based bio-printing. Biomaterials. 2016;76:321–43.
4. Pati F, Ha D-H, Jang J, Han HH, Rhie J-W, Cho D-W. Biomimetic 3D tissue printing for soft tissue regeneration. Biomaterials. 2015;62:164–75.
5. Shim J-H, Lee J-S, Kim JY, Cho D-W. Bioprinting of a mechanically enhanced three-dimensional dual cell-laden construct for osteochondral tissue engineering using a multi-head tissue/organ building system. J Micromech Microeng. 2012;22:085014.
6. Lee J-S, Kim BS, Seo D, Park JH, Cho D-W. Three-dimensional cell printing of large-volume tissues: application to ear regeneration. Tissue Eng Part C Methods. 2017;23:136–45.
7. Mandrycky C, Wang Z, Kim K, Kim D-H. 3D bioprinting for engineering complex tissues. Biotechnol Adv. 2016;34:422–34.
8. Murphy SV, Atala A. 3D bioprinting of tissues and organs. Nat Biotechnol. 2014;32:773–85.
9. Guillotin B, Souquet A, Catros S, Duocastella M, Pippenger B, Bellance S, et al. Laser assisted bioprinting of engineered tissue with high cell density and microscale organization. Biomaterials. 2010;31:7250–6.
10. Koch L, Kuhn S, Sorg H, Gruene M, Schlie S, Gaebel R, et al. Laser printing of skin cells and human stem cells. Tissue Eng Part C Methods. 2009;16:847–54.
11. Guillemot F, Souquet A, Catros S, Guillotin B, Lopez J, Faucon M, et al. High-throughput laser printing of cells and biomaterials for tissue engineering. Acta Biomater. 2010;6:2494–500.
12. Kim JD, Choi JS, Kim BS, Choi YC, Cho YW. Piezoelectric inkjet printing of polymers: stem cell patterning on polymer substrates. Polymer. 2010;51:2147–54.
13. Chang CC, Boland ED, Williams SK, Hoying JB. Direct-write bioprinting three-dimensional biohybrid systems for future regenerative therapies. J Biomed Mater Res B Appl Biomater. 2011;98:160–70.

14. Michael S, Sorg H, Peck C-T, Koch L, Deiwick A, Chichkov B, et al. Tissue engineered skin substitutes created by laser-assisted bioprinting form skin-like structures in the dorsal skin fold chamber in mice. PLoS One. 2013;8:e57741.
15. Norotte C, Marga FS, Niklason LE, Forgacs G. Scaffold-free vascular tissue engineering using bioprinting. Biomaterials. 2009;30:5910–7.
16. Smith CM, Stone AL, Parkhill RL, Stewart RL, Simpkins MW, Kachurin AM, et al. Three-dimensional bioassembly tool for generating viable tissue-engineered constructs. Tissue Eng. 2004;10:1566–76.
17. Xu T, Gregory CA, Molnar P, Cui X, Jalota S, Bhaduri SB, et al. Viability and electro-physiology of neural cell structures generated by the inkjet printing method. Biomaterials. 2006;27:3580–8.
18. Chang R, Nam J, Sun W. Effects of dispensing pressure and nozzle diameter on cell survival from solid freeform fabrication–based direct cell writing. Tissue Eng Part A. 2008;14:41–8.
19. Mironov V, Kasyanov V, Markwald RR. Organ printing: from bioprinter to organ biofabrica-tion line. Curr Opin Biotechnol. 2011;22:667–73.
20. Marga F, Jakab K, Khatiwala C, Shepherd B, Dorfman S, Hubbard B, et al. Toward engineer-ing functional organ modules by additive manufacturing. Biofabrication. 2012;4:022001.
21. Jones N. Science in three dimensions: the print revolution. Nat News. 2012;487:22.

Chapter 5
Conventional Bioinks

5.1 Overview of Bioinks

Bioink is a printable hydrogel that is able to encapsulate cells for fabrication of living structures when applied to a 3D bioprinting system [1]. According to the origin of sources, it is classified into natural and synthetic bioinks [2, 3]. It is a natural bioink if the material is derived from nature, such as alginate or collagen. On the other hand, a synthetic bioink is made from synthetic materials such as polyethylene glycol (PEG) or pluronic F127. Though these two sources are fundamentally different, they should satisfy common criteria to be used as a bioink. Most importantly, biocompatibility is essential so that the viability of the cells in the bioinks can be guaranteed. Besides, the shear thinning property, by which the viscosity decreases when the pressure is applied, is required to ensure the cell viability during the extrusion via micro-nozzle, especially for the extrusion printing system [3]. Moreover, proper mechanical and crosslinking properties are needed for shape fidelity and structural maintenance. A variety of bioinks have been reported so far to achieve the abovementioned properties [1–3].

5.2 Natural Bioinks

Natural bioinks are made from natural sources and are safe to apply to printing cells because of their low toxicity [4]. Some natural bioinks, however, are not well suited for cell function or growth, while others are functionally good but may not be as good in terms of printability. As a result, many attempts have been made to overcome the inherent disadvantages of certain natural bioinks by using bioinks made of a mixture of natural materials rather than a single substance or chemically modified

© Springer Nature Switzerland AG 2019
D.-W. Cho et al., *3D Bioprinting*, https://doi.org/10.1007/978-3-030-32222-9_5

ones. In this regard, this section will cover the range of natural bioinks, including origin of raw material, chemical or biological properties, printability, crosslinking properties, etc.

5.2.1 Alginate

Alginate, also known as algin or alginic acid, a natural substance derived from brown algae, is an anionic polysaccharide with a chemical structure in which mannuronic and guluronic acids are repeated [5]. Alginate is a cheap and easily obtainable material and has a water-soluble property suitable for preparing hydrogel, although the rate of dissolution in water is low. When the alginate hydrogel is treated with the calcium chloride, the calcium cation and the anion backbone of the alginate bind to each other, and the crosslinking is immediately carried out, resulting in gel. Alginate has been traditionally used in tissue engineering and regenerative medicine, because it shows no cytotoxicity and does not induce inflammation or immune response in vivo. Therefore, due to its crosslinking properties along with safety, it has been widely used as a bioink in 3D cell printing technology for fabrication of complex living structures [6]. However, since the chemical structure of alginate does not have motifs for cell adhesion, it is difficult to expect proliferations or matured functions in cell works using alginate [7]. In order to overcome the bioinert property, conjugation of cell-attachable peptides, such as Arg-Gly-Asp (RGD) derived from fibronectin or Asp-Gly-Glu-Ala (DGEA) derived from collagen, has been reported [8]. Since toxic substances are involved in these chemical modifications and it takes time until the final use, most of them are used in combination with other materials that can attach cells but have poor mechanical properties, rather than using alginate alone or chemically modified alginate. Yu et al. fabricated cartilage progenitor cell-laden 3D vascular structure, and it showed high cell viability (~90%) until 72-h culture [9]. Besides, Chang et al. encapsulated HepG2 hepatoma cell line to alginate bioink to print perfusable 3D liver structure, and efficiency of drug conversion was improved in the perfusion 3D condition [10]. In the case of complicated tissue, Shim et al. fabricated a 3D osteochondral by 3D printing the polycaprolactone (PCL) framework and two kinds of cell-laden alginate bioinks where chondrocyte and osteoblast were encapsulated, respectively, and high cell viability was observed until 7 days (Fig. 5.1a) [11].

5.2.2 Agarose

Agarose is a polysaccharide separated from red seaweed, which has a repeated chemical structure of agarobiose composed of galactose and anhydro-L-galactopyranose [12]. It is a traditionally used material in electrophoresis in molecular biology research, and it shows a reversible gelation property by which it

Fig. 5.1 Natural bioinks printed 3D constructs. (**a**) Complicated osteochondral construct fabricated using alginate bioink and PCL framework [11]. (**b**) 3D printing of tissue spheroid filaments using agarose bioink. (Reprinted from [15], Copyright (2009), with permission from Elsevier). (**c**) Stem cell-laden chondrogenic structure using chitosan bioink [19]. (**d**) 3D-printed scaffold using silk fibroin bioink. (Copyright (2008) Wiley. Used with permission from [22]). (**e**) 3D cell-printed structure using collagen bioink and crosslinked by genipin. (Reproduced with permission from [28]). (**f**) 3D-printed gelatin scaffold. (Reprinted from [45], Copyright (2018), with permission from Elsevier). (**g**) 3D-printed and UV-crosslinked methacrylated HA bioink [46]. (**h**) 3D-printed HUVEC-laden microvascular structure using fibrin bioink. (Reprinted from [47], Copyright (2009), with permission from Elsevier). (**i**) 3D hepatocyte printed structure using Matrigel bioink. (Reproduced with permission from [43])

solidifies at low temperatures and becomes solution at 20–70 °C. As with alginate, though it shows good biocompatibility, it is not a proper material for cell culture because there is no cell-binding motif [13]. However, in terms of fabrication, it is suitable for 3D printing because it is easy to control its physical properties and its shape can be maintained for a certain time when gelation occurs. Moreover, agarose is used as a sacrificial material to form desired void spaces or hollow channels inside bulky structures due to its thermal-reversible advantage [14]. Norotte et al. encapsulated agarose bioink with human umbilical vein smooth muscle cells (HUVSMC) and human skin fibroblast (HSF), respectively, to produce the cylinder construct of the controllable diameter from 300 to 500 μm, which showed that the multicellular patterned construct could be developed (Fig. 5.1b) [15].

5.2.3   Chitosan

Chitosan is a cationic amino-polysaccharide obtained by deacetylation of chitin isolated from shells of crustaceans, insects, or fungi and is widely used in many fields such as pharmaceutical, medical, and biotechnology due to its biocompatible, assessable, and controllable properties [16]. In addition, it is known that the chemical structure of glycosaminoglycan is similar to that of the human body, and it is also used in the tissue engineering field due to its advantages such as promoting wound healing in terms of functionality [17]. Chitosan is a weak basic substance, which can be dissolved in a weak acid and made into a hydrogel, and it undergoes crosslinking by multivalent anion such as phosphate from tripolyphosphate (TPP). In addition, when mixed with anionic material such as alginate or chondroitin sulfate, it forms a polyelectrolyte complex and becomes crosslinked [18]. Chitosan is much better in terms of cell functionality than alginate or agarose, but it is used in combination with alginate and other materials, rather than being used alone, because 3D printing does not have printability sufficient to make elaborate structures [19]. Ye et al. printed a stem cell-laden construct using bioink chitosan, and chondrogenic differentiation was induced to prepare cartilage-like structure (Fig. 5.1c). After 4 weeks, chondrogenic makers such as type II collagen and aggrecan were highly expressed in gene and protein levels, showing successful preparation of osteochondral graft.

5.2.4   Silk Fibroin

Silk fibroin is a natural polymer remaining after removal of sericin from silk produced by silkworm or spider. It is a very widely used material in the biomedical field due to its favorable mechanical properties, controllability, biodegradability, and biocompatibility [20]. The chemical structure of fibroin is composed of a stabilized β-sheet structure due to hydrogen bonding and hydrophobic interaction, and it has a sol-gel transition when this hydrogen bonding is broken through dialysis to make it water-soluble. Gelation of fibroin hydrogel can be achieved by various methods such as heating, sonication, and chemical treatment, and photo-crosslinking through chemical modification is also available, as with other materials such as gelatin [21]. Ghosh et al. fabricated 3D microperiodic scaffold using silk fibroin bioink, and functionality of the scaffold was confirmed by adhesion, proliferation, and chondrogenic differentiation capability of mesenchymal stem cells cultured on it (Fig. 5.1d) [22].

5.2.5 Collagen

Collagen is the most abundant structural protein in our body, and it can be derived from various sources, such as human, pig, rat, and mouse [23]. It shows good performance in cell attachment, proliferation, and function by its integrin-binding domains, and it is the most used material in biomedical fields due to its processability, accessibility, and biocompatibility [24]. Collagen is an insoluble protein and requires an acidic solvent to solubilize it to make it a hydrogel. In order to be used as a bioink capable of encapsulating cells, it should be neutralized at a temperature as low as ~4 °C, because it undergoes thermal crosslinking at high temperature (~37 °C), which is similar to physiological condition [25]. However, since collagen is not a material with a high stiffness and its crosslinking process takes a long time, there are limitations to fabricating elaborate or bulky structures using collagen bioink in 3D printing technology. To overcome this problem, studies have been carried out to support collagen-printed structure by framework using synthetic polymers such as PCL [26]. Lee et al. used a collagen bioink to construct a liver construct, and collagen bioink containing hepatocyte, HUVEC, and lung fibroblast was printed in the PCL framework, showing possibility of collagen-based complicated living constructs with predesigned porous structures. In the study by Diamantides et al., mechanical properties of collagen bioink were enhanced by mixing the natural photo-crosslinker riboflavin (vitamin B2) upon irradiation with UV [27]. Moreover, there have also been studies to improve stiffness after crosslinking by additional crosslinking using genipin crosslinker, which is known to be less toxic than aldehyde-based chemical crosslinkers (Fig. 5.1e) [28].

5.2.6 Gelatin

Gelatin is a biomaterial produced by irreversible thermal denaturation of collagen, and it also can be produced by acidic or basic processing of collagen [25]. There are many types of collagen, but the only source of gelatin is type I collagen, which has no cysteine sequence. While the collagen is insoluble protein, gelatin can be dissolved in water over 30 °C. Besides, its gelation condition is opposite to that of collagen, and it undergoes gelation at 4 °C and becomes solution at 37 °C. Gelatin is a suitable material for biomedical research because of its processability, accessibility, and minimal immune response, and it is also very well-known as a bioink in the 3D printing field. Since the gelation of gelatin is reversible, additional crosslinking must be performed using a chemical crosslinker such as aldehyde derivatives or genipin after constructing the structure by 3D printing (Fig. 5.1f) [29]. In addition, gelatin methacrylate (GelMA), produced by reaction of methacrylic anhydride with gelatin, is a very popular photocurable gelatin, which can improve the both cross-linking and mechanical properties [30]. GelMA is often used alone to fabricate sophisticated structures and is also considered as an additive for various bioinks for enhancement of printability and mechanical properties. Bertassoni et al. fabricated HepG2 hepatoma cell- and NIH-3T3 fibroblast-laden 3D structure using GelMA bioink, and cell viability was maintained at more than 80% after culturing for 8 days [31].

5.2.7   Hyaluronic Acid

Hyaluronic acid (HA), or hyaluronan, is an anionic non-sulfated glycosaminogly-can, and it consists of repeated disaccharide unit of glucuronic acid and acetylglu-cosamine [32]. HA is distributed in most connective tissues and is predominantly present in ECM of cartilage and brain. HA is a biocompatible and biodegradable material that is widely used in delivering bioactive molecules or clinical applica-tions for joints and arthritis [33]. Moreover, since the structure of HA is identical in all mammals, it has the advantage that it is not recognized as nonself in recipient, regardless of its origin [34]. With respect to biodegradability, it does not induce any inflammatory response because it is completely decomposed into water and carbon dioxide in the body. Though HA shows very good solubility to prepare hydrogel, the mechanical properties of HA hydrogel is poor. Therefore, an additional crosslinking process is required to fabricate the desired shape using 3D printing. To this end, attempts have been made to use a chemical crosslinker such as divinyl sulfone (DVS) and butanediol diglycidyl ether (BDDE), and, like GelMA where meth-acrylic acid is conjugated, HA also can be modified to be a photocurable material in the same functional group (Fig. 5.1g) [35, 36]. As an example of using HA bioink, Park et al. printed osteoblast-laden HA bioink, and the osteogenic marker expressions were found to be higher than those of the structure using type I collagen bioink [37].

5.2.8   Fibrin

Fibrin is produced by the enzymatic reaction of fibrinogen and thrombin involved in blood coagulation, and it plays an important role in cell survival and proliferation [38]. Fibrinogen, a precursor of fibrin, is a glycoprotein and is converted to fibrin with help of thrombin and calcium ion. Fibrin gel is often used to replace damaged tissue because of its biodegradable and biocompatible properties, with advantages in wound healing. When applied to 3D printing, it can be printed alone, and it is used in combination with other materials such as collagen or alginate to improve mechanical properties or printability (Fig. 5.1h). Xu et al. fabricated a hepatic structure using hepatic cell-encapsulated gelatin/fibrin bioink, and it was cross-linked with thrombin [39]. In the fabricated structure, glutamate-oxaloacetate trans-aminase (GOT) and albumin secretion were observed until day 13, demonstrating the development of a functional construct for medical regeneration.

5.2.9 ECM Complex

The abovementioned bioinks are mostly derived from a part of the ECM components, excluding the cases of some algae. In organisms, ECM is not a single substance, but it is a complex consisting of different types of biomolecules including collagen, fibronectin, laminin, growth factor, and cytokine [40]. Thus, rather than using only bioinks made of a single component of ECM, studies have been done on the application of complicated ECMs. For example, Matrigel is a hydrogel of basement membrane derived from Engelbreth-Holm-Swarm (EHS) mouse sarcoma cells, and it has complicated ECM components, such as type IV collagen or fibronectin [41]. Since collagen is the most abundant in the Matrigel, crosslinking conditions, mechanical properties, and printability are very similar to those of collagen [42]. Because Matrigel shows better performance than other singular materials, it is widely used in studies to develop in vitro models. However, it is rarely used in regenerative medicine because it is derived from mouse cancer cells. As an example using Matrigel bioink, Snyder et al. 3D-printed a multicellular hepatic construct demonstrating biomimetic radioprotection by hepatic prodrug conversion metabolism (Fig. 5.1i) [43]. Recently, to reproduce the biochemical microenvironment of native tissue/organ, there have been studies developing bioinks using residual ECM after decellularizing target tissue/organ [44]. This is called decellularized extracellular matrix-derived bioink (dECM bioink), and it will be covered intensively in the next chapter.

5.3 Synthetic Bioinks

Other than natural bioinks, synthetic bioinks are also a class of printable hydrogel that is made from chemically synthesized material. They enable the fabrication of delicate or large volumetric constructs because of their tunable mechanical properties and crosslinking manners [48]. However, due to the issue of material sources, the synthetic bioinks are tackled in the recapitulation of actual ECM environments of native tissue/organ when compared to natural bioinks [49]. As a result, the synthetic bioinks generally show inferior performances of promoting the behaviors of cells such as survival, adhesion, and maturation. Moreover, the cytotoxicity of synthetic biomaterials impairs cell viability and function, limiting a long-term maintenance of in vitro models. For these reasons, only few of them have been applied for modeling in vitro tissues/organs. Taken together, this chapter mainly focuses on the naturally derived bioinks rather than the synthetic bioinks; instead, detailed information of synthetic bioinks is summarized in related articles [4].

References

1. Groll J, Burdick J, Cho D, Derby B, Gelinsky M, Heilshorn S, Jüngst T, Malda J, Mironov V, Nakayama K. A definition of bioinks and their distinction from biomaterial inks. Biofabrication. 2018;11(1):013001.
2. Jang J. 3D bioprinting and in vitro cardiovascular tissue modeling. Bioengineering. 2017;4(3):71.
3. Jang J, Park JY, Gao G, Cho D-W. Biomaterials-based 3D cell printing for next-generation therapeutics and diagnostics. Biomaterials. 2018;156:88–106.
4. Gungor-Ozkerim PS, Inci I, Zhang YS, Khademhosseini A, Dokmeci MR. Bioinks for 3D bioprinting: an overview. Biomater Sci. 2018;6(5):915–46.
5. Kim HS, Lee C-G, Lee EY. Alginate lyase: structure, property, and application. Biotechnol Bioprocess Eng. 2011;16(5):843.
6. Kundu J, Shim JH, Jang J, Kim SW, Cho DW. An additive manufacturing-based PCL–alginate–chondrocyte bioprinted scaffold for cartilage tissue engineering. J Tissue Eng Regen Med. 2015;9(11):1286–97.
7. Murphy WL, Mercurius KO, Koide S, Mrksich M. Substrates for cell adhesion prepared via active site-directed immobilization of a protein domain. Langmuir. 2004;20(4):1026–30.
8. Faulkner-Jones A, Fyfe C, Cornelissen D-J, Gardner J, King J, Courtney A, Shu W. Bioprinting of human pluripotent stem cells and their directed differentiation into hepatocyte-like cells for the generation of mini-livers in 3D. Biofabrication. 2015;7(4):044102.
9. Yu Y, Zhang Y, Martin JA, Ozbolat IT. Evaluation of cell viability and functionality in vessel-like bioprintable cell-laden tubular channels. J Biomech Eng. 2013;135(9):091011.
10. Chang R, Emami K, Wu H, Sun W. Biofabrication of a three-dimensional liver micro-organ as an in vitro drug metabolism model. Biofabrication. 2010;2(4):045004.
11. Shim J-H, Lee J-S, Kim JY, Cho D-W. Bioprinting of a mechanically enhanced three-dimensional dual cell-laden construct for osteochondral tissue engineering using a multi-head tissue/organ building system. J Micromech Microeng. 2012;22(8):085014.
12. Venkatesan J, Anil S, Kim S-K. Introduction to seaweed polysaccharides. Seaweed polysaccharides. Amsterdam: Elsevier; 2017. p. 1–9.
13. Tanaka N, Moriguchi H, Sato A, Kawai T, Shimba K, Jimbo Y, Tanaka Y. Microcasting with agarose gel via degassed polydimethylsiloxane molds for repellency-guided cell patterning. RSC Adv. 2016;6(60):54754–62.
14. Zhang YS, Yue K, Aleman J, Mollazadeh-Moghaddam K, Bakht SM, Yang J, Jia W, Dell'Erba V, Assawes P, Shin SR. 3D bioprinting for tissue and organ fabrication. Ann Biomed Eng. 2017;45(1):148–63.
15. Norotte C, Marga FS, Niklason LE, Forgacs G. Scaffold-free vascular tissue engineering using bioprinting. Biomaterials. 2009;30(30):5910–7.
16. Kumar MNR. A review of chitin and chitosan applications. React Funct Polym. 2000;46(1):1–27.
17. Katalinich M. Characterization of chitosan films for cell culture applications. PhD Thesis. 2001.
18. Ahmadi F, Oveisi Z, Samani SM, Amoozgar Z. Chitosan based hydrogels: characteristics and pharmaceutical applications. Res Pharmaceu Sci. 2015;10(1):1.
19. Ye K, Felimban R, Traianedes K, Moulton SE, Wallace GG, Chung J, Quigley A, Choong PF, Myers DE. Chondrogenesis of infrapatellar fat pad derived adipose stem cells in 3D printed chitosan scaffold. PLoS One. 2014;9(6):e99410.

20. Kundu B, Rajkhowa R, Kundu SC, Wang X. Silk fibroin biomaterials for tissue regenerations. Adv Drug Deliv Rev. 2013;65(4):457–70.
21. Qi Y, Wang H, Wei K, Yang Y, Zheng R-Y, Kim I, Zhang K-Q. A review of structure construction of silk fibroin biomaterials from single structures to multi-level structures. Int J Mol Sci. 2017;18(3):237.
22. Ghosh S, Parker ST, Wang X, Kaplan DL, Lewis JA. Direct-write assembly of microperiodic silk fibroin scaffolds for tissue engineering applications. Adv Funct Mater. 2008;18(13):1883–9.
23. Cho D-W, Lee H, Han W, Choi Y-J. Bioprinting of liver. 3D bioprinting in regenerative engineering: principles and applications. Boca Raton, FL: CRC Press; 2018.
24. Sionkowska A, Skrzyński S, Śmiechowski K, Kołodziejczak A. The review of versatile application of collagen. Polym Adv Technol. 2017;28(1):4–9.
25. Gorgieva S, Kokol V. Collagen-vs. gelatine-based biomaterials and their biocompatibility: review and perspectives. Biomaterials applications for nanomedicine. Rijeka: IntechOpen; 2011.
26. Lee JW, Choi Y-J, Yong W-J, Pati F, Shim J-H, Kang KS, Kang I-H, Park J, Cho D-W. Development of a 3D cell printed construct considering angiogenesis for liver tissue engineering. Biofabrication. 2016;8(1):015007.
27. Diamantides N, Wang L, Pruiksma T, Siemiatkoski J, Dugopolski C, Shortkroff S, Kennedy S, Bonassar LJ. Correlating rheological properties and printability of collagen bioinks: the effects of riboflavin photocrosslinking and pH. Biofabrication. 2017;9(3):034102.
28. Kim YB, Lee H, Kim GH. Strategy to achieve highly porous/biocompatible macroscale cell blocks, using a collagen/genipin-bioink and an optimal 3D printing process. ACS Appl Mater Interfaces. 2016;8(47):32230–40.
29. Wang X, Ao Q, Tian X, Fan J, Tong H, Hou W, Bai S. Gelatin-based hydrogels for organ 3D bioprinting. Polymers. 2017;9(9):401.
30. Billiet T, Gevaert E, De Schryver T, Cornelissen M, Dubruel P. The 3D printing of gelatin methacrylamide cell-laden tissue-engineered constructs with high cell viability. Biomaterials. 2014;35(1):49–62.
31. Bertassoni LE, Cardoso JC, Manoharan V, Cristino AL, Bhise NS, Araujo WA, Zorlutuna P, Vrana NE, Ghaemmaghami AM, Dokmeci MR. Direct-write bioprinting of cell-laden methacrylated gelatin hydrogels. Biofabrication. 2014;6(2):024105.
32. Coleman SR, Committee PSEFD. Cross-linked hyaluronic acid fillers. Plast Reconstr Surg. 2006;117(2):661–5.
33. Khunmanee S, Jeong Y, Park H. Crosslinking method of hyaluronic-based hydrogel for biomedical applications. J Tiss Eng. 2017;8:2041731417726464.
34. Naoum C, Dasiou-Plakida D. Dermal filler materials and botulin toxin. Int J Dermatol. 2001;40(10):609–21.
35. Maiz-Fernández S, Pérez-Álvarez L, Ruiz-Rubio L, Pérez González R, Sáez-Martínez V, Ruiz Pérez J, Vilas-Vilela JL. Synthesis and characterization of covalently crosslinked pH-responsive hyaluronic acid nanogels: effect of synthesis parameters. Polymers. 2019;11(4):742.
36. Möller S, Weisser J, Bischoff S, Schnabelrauch M. Dextran and hyaluronan methacrylate based hydrogels as matrices for soft tissue reconstruction. Biomol Eng. 2007;24(5):496–504.
37. Park JY, Choi J-C, Shim J-H, Lee J-S, Park H, Kim SW, Doh J, Cho D-W. A comparative study on collagen type I and hyaluronic acid dependent cell behavior for osteochondral tissue bioprinting. Biofabrication. 2014;6(3):035004.
38. Rajangam T, An SSA. Fibrinogen and fibrin based micro and nano scaffolds incorporated with drugs, proteins, cells and genes for therapeutic biomedical applications. Int J Nanomedicine. 2013;8:3641.
39. Xu W, Wang X, Yan Y, Zheng W, Xiong Z, Lin F, Wu R, Zhang R. Rapid prototyping three-dimensional cell/gelatin/fibrinogen constructs for medical regeneration. J Bioact Compat Polym. 2007;22(4):363–77.
40. Hoshiba T, Lu H, Kawazoe N, Chen G. Decellularized matrices for tissue engineering. Expert Opin Biol Ther. 2010;10(12):1717–28.

41. Kleinman HK, Martin GR. Matrigel: basement membrane matrix with biological activity. Semin Cancer Biol. 2005;15:378–86.
42. Kloxin AM, Kloxin CJ, Bowman CN, Anseth KS. Mechanical properties of cellularly responsive hydrogels and their experimental determination. Adv Mater. 2010;22(31):3484–94.
43. Snyder J, Hamid Q, Wang C, Chang R, Emami K, Wu H, Sun W. Bioprinting cell-laden matrigel for radioprotection study of liver by pro-drug conversion in a dual-tissue microfluidic chip. Biofabrication. 2011;3(3):034112.
44. Pati F, Jang J, Ha D-H, Kim SW, Rhie J-W, Shim J-H, Kim D-H, Cho D-W. Printing three-dimensional tissue analogues with decellularized extracellular matrix bioink. Nat Commun. 2014;5:3935.
45. Lewis PL, Green RM, Shah RN. 3D-printed gelatin scaffolds of differing pore geometry modulate hepatocyte function and gene expression. Acta Biomater. 2018;69:63–70.
46. Poldervaart MT, Goversen B, De Ruijter M, Abbadessa A, Melchels FP, Öner FC, Dhert WJ, Vermonden T, Alblas J. 3D bioprinting of methacrylated hyaluronic acid (MeHA) hydrogel with intrinsic osteogenicity. PLoS One. 2017;12(6):e0177628.
47. Cui X, Boland T. Human microvasculature fabrication using thermal inkjet printing technology. Biomaterials. 2009;30(31):6221–7.
48. Guvendiren M, Burdick JA. Engineering synthetic hydrogel microenvironments to instruct stem cells. Curr Opin Biotechnol. 2013;24(5):841–6.
49. Donderwinkel I, Van Hest JC, Cameron NR. Bio-inks for 3D bioprinting: recent advances and future prospects. Polym Chem. 2017;8(31):4451–71.

Chapter 6
Decellularized Extracellular Matrix-Based Bioinks

6.1 Definition and Advantages of dECM Bioink

Decellularized extracellular matrix (dECM) is a form of acellular tissue where the cellular components are removed on purpose, and it contains plenty of functional materials inherited from native tissue [1]. In living organisms, each tissue/organ is composed of its specific composition of structural ECMs, such as various types of collagen, fibronectin, elastin, laminin, glycosaminoglycans, and hyaluronic acid, and those components are essential to maintain architecture and functions of the tissues. Besides, various functional factors including growth factors, cytokines, and enzymes in ECM also play a critical role for cell proliferation, differentiation, and parenchymal function in tissue/organ [2]. Therefore, in tissue engineering and regenerative medicine, utilizing dECM enables benefits of repairing damaged tissues and recapitulating the optimized environments for tissue cells. Since its application was first reported in 1973 for wound dressings, dECM has been used for many studies and clinical trials in various forms such as acellular tissue shaped-scaffold, lyophilized sponge, injectable hydrogel, or coating material for cell culture [1, 3].

A dECM bioink, which is defined as a tissue-specific bioink, is a solubilized form of dECM for application in a 3D printing system. As requirements for bioink compared with other types of dECM material, it is necessary to satisfy the capability of cell encapsulation, rheological properties suitable for extrusion via microneedle, and proper stiffness and crosslinking properties for shape fidelity [4]. Combining with the 3D printing technology, dECM bioink has considerable advantages for development of artificial tissues in terms of recapitulating both structures and biochemical environments of native tissues. Accordingly, not only the field of regenerative medicine but also the studies that develop in vitro tissue/organ models have been using dECM bioink extensively [5].

© Springer Nature Switzerland AG 2019
D.-W. Cho et al., *3D Bioprinting*, https://doi.org/10.1007/978-3-030-32222-9_6

6.2   Preparation of dECM Bioinks

6.2.1   Origin of the Tissue

Considering the origin of dECM, mouse, rat, rabbit, goat, bovine, porcine, and human are reported as sources; however, among them, human-derived tissue itself is the most suitable for recapitulating the human tissue environment. However, compared with other species, accessibility of human tissues is not appropriate to ensure sufficient amounts [5]. Instead, it has been reported that porcine are considered to be the most proper source, because body mass and physiological response of porcine is similar to human [6]. Besides, they are ready to be utilized in terms of easy availability and sustainability. Moreover, porcine shows the most homogeneity with human among other species, except nonhuman primates. For example, in the case of transforming growth factor beta (TGFβ), which is essential for tissue regeneration and development, it is genetically the same in human and pig [7]. Therefore, most studies have reported to prepare dECM bioink using porcine tissues. Although porcine tissue is a good source for dECM, xenoantigens such as swine leukocyte antigen (SLA) class II and galactose α-1,3 galactose (α-Gal) can induce unexpected responses in immune cells or stem cells, and the risk due to interspecies differences cannot be completely excluded [8–10].

6.2.2   Decellularization

The purpose of decellularization is to remove cell elements within the tissue and to preserve extracellular matrix (ECM) as much as possible. To this end, several methods have been developed to decellularize tissues in various ways, and descriptions in detail are shown in the table (Table 6.1). Generally, processes for decellularization include mechanical, chemical, and enzymatic methods, and, in most cases, decellularization is carried out by the process in which they are combined [11]. In the case of mechanical methods, tissue/organ can be made into small or thin pieces via chopping or slicing in order to maximize the efficiency of the decellularizing agents to be treated. Freezing–thawing cycle also can damage tissues facilitating decellularization process. The chemical process involves treatment with detergents to break cellular membrane consisting of the lipid bilayer, and it also washes out cell debris from decellularizing tissues. Among detergents, ionic detergents such as sodium dodecyl sulfate SDS are known to provide harsher conditions than nonionic ones, including triton X-100 [12]. The former is excellent for cell removal, but it also removes the ECM component, while the latter is suitable for ECM preservation, but it is not excellent for cell component removal compared to the former. Therefore, balancing them is required to achieve optimized efficiency of decellularization. Other than detergents, osmotic stress lysing cells is also available using hypertonic and hypotonic solutions. With respect to the enzymatic process, trypsin

Table 6.1 Methods of decellularization

Category	Mechanism	Agents and methods	Description
Mechanical process	– Cell lysis by mechanical stress – Increase in surface area for decellularization agents	Freezing–thawing cycle	Tissue and cell lysis
		Chopping, mincing, or slicing	– Breaking down tissue structures – Increase of surface area for other reagents
Chemical process	– Disruption of plasma membrane lysing cells – Washing out cell debris	Triton X-100	Nonionic detergent
		Sodium dodecyl sulfate (SDS)	Ionic detergent
		Sodium deoxycholic acid (SDA)	Ionic detergent
		NaCl	Hypertonic solution
		Ammonium hydroxide	Basic solution
		Ethylenediaminetetraacetic acid (EDTA)	Chelating agent
		Isopropyl alcohol	Removal of lipid
		Peracetic acid/ethanol	Sterilization
		3-[(3-Chloamidopropyl)dimethyl-ammonio]-1-propanesulfonate (CHAPS)	Zwitterionic detergent
Enzymatic process	– Increasing of porosity and loosening of tissue – Degradation of nucleic acid	Trypsin	Protease activity
		DNase	Degradation of DNA
		RNase	Degradation of RNA

cuts cell–matrix interaction and increases the efficiency of chemical treatment by making it porous through the cleavage ECM network. Nucleases such as DNase or RNase degrade DNA or RNA components, which can induce unexpected inflammation. Note that most of the chemicals or enzymes used during the process are cytotoxic, and if these are not removed properly, dECM bioink itself will have a detrimental effect on the cells. Therefore, it is very important to wash thoroughly between treatment steps using sterile distilled water or PBS.

6.2.3 Validation of Decellularization

In decellularization, it is ideal to remove the cell components that unexpectedly cause harmful effects such as inflammations or immune responses, while preserving the ECM components that benefit the behaviors of the cells. To address this, various biochemical assays or staining methods have been reported to evaluate efficiency of decellularization, and confirming remnant DNA, glycosaminoglycan (GAG), and collagen is a widely accepted protocol (Fig. 6.1a).

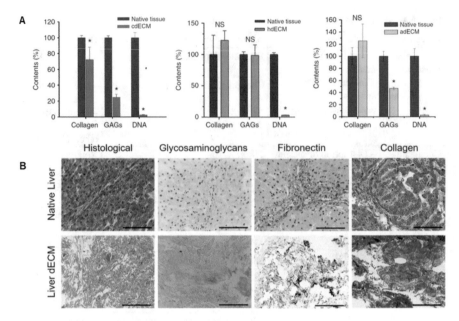

Fig. 6.1 Validation of decellularization process. (**a**) Quantification of glycosaminoglycans (GAGs), collagen, and DNA component in cartilage dECM (cdECM), heart dECM (hdECM), and adipose dECM (adECM), respectively, compared with their native tissues. (Reproduced with permission from [4]). (**b**) Histological analysis of nucleus, GAGs, fibronectin, and collagen in liver dECM compared to its native tissue. (Reproduced with permission from [18]. Copyright 2017 American Chemical Society)

In the case of residual DNA in the freeze-dried dECM, which is considered a representative of cellular components, it is known to have no harmful effect if the amount is less than 3% of freeze-dried native tissue, less than 50 ng in a mg of dECM, and less than 200 base pairs [4, 11]. To confirm remaining dsDNA, DNA binding reagents such as Hoechst 33258, diamidino-phenylindole (DAPI), or PicoGreen can be used, and measuring absorbance at 260 nm is also available, though it is not very precise compared with methods using those reagents [13, 14]. The amount of remaining GAG can be confirmed via various methods, among which dimethylmethylene blue (DMMB) assay is based on the principle that absorbance shifts when DMMB is bound to GAG [15]. It can also be confirmed histologically through alcian blue staining, which stains the GAG region in blue. For collagen, the quantifying method relies on hydroxyproline, which is a certain kind of amino acid exclusively observed in collagen [16]. Besides, it can be shown via picrosirius red or Masson's trichrome staining, which stain collagen in red and blue, respectively. In the cases of both GAG and collagen, there have been no experimental results showing optimized amount of each component, but both are crucial ECM components affecting viability and functions of cell. Therefore, the greater the amount of each component, the better the expected function. In addition to these, important elements such as fibronectin, laminin, or elastin can be also identified using assays utilizing a specific antibody for each component according to the interest of the researcher (Fig. 6.1b) [17, 18].

Recently, in the case of remaining proteins in particular after decellularization, whole proteins can be profiled by mass spectrometry, instead of targeting only a few of them. This analytic approach enables to discuss the matrisome, which is a group of proteins in the ECM participating in actual signaling to cells, yielding insight into how they will affect target cells [19]. Moreover, damage-associated molecular patterns (DAMPs), which were beyond criteria for decellularization, such as heat shock proteins, high-mobility group box 1 (HMGB1), and histone protein, were also observed through this technique [20]. To sum up, the technique suggested the need for further study of the standard of decellularization, allowing for a discussion of the matrisome and DAMPs, which are potent factors for tissue specificity and unexpected immune responses, respectively.

6.2.4 Solubilization and Neutralization

To prepare bioink, lyophilized dECM tissue should be solubilized to become hydrogel. Collagens are the most abundant component in dECM, and they are also well-known insoluble proteins in neutral pH [21]. Therefore, acidic solvents are necessary to solubilize dECM. The collagen components in dECM contain intact telopeptides, which lower solubility and have immunogenicity [22]. Therefore, to prepare freeze-dried dECM as a hydrogel, both acidic solvent and removal of telopeptides are essential. Conventionally, 0.5 M acetic acid and 0.1 M hydrochloride are the most commonly used acidic solutions [23]. To remove the telopeptides, pepsin is a widely used protease showing its enzymatic activity in the acidic condition, and it selectively degrades telopeptides of collagen [24]. Solubilization is usually carried out using a stirrer until dECM is completely dissolved, and if there are any remaining particles, they are filtered to prevent the nozzle from clogging during printing. To finally prepare a bioink capable of encapsulating the living cells, the neutralization process is essential. During neutralization, 10N sodium hydroxide (NaOH) solution is widely used for titration, and it is important to keep it at a low temperature to prevent unintended thermal crosslinking of the dECM bioink prior to cell encapsulation [4].

6.3 Rheological Properties of dECM Bioinks

Rheological properties of dECM bioink are closely related to the applicability to 3D cell printing, because viability of the cells and shape fidelity of cell-printed construct are determined by them. For the extruding process, viscosity of the cell-laden bioink should be lowered to secure high cell viability, and it is necessary to recover it as soon as possible after extrusion to maintain the structure of interest. These are called shear thinning and thixotropic properties, respectively, and both are essential criteria for bioink in 3D cell printing research [25, 26]. Besides, to maintain the cell-laden structure for using downstream applications, stiffness should be secured, and it is attained by temperature-dependent crosslinking properties of dECM bioink [27]. As in collagen, dECM bioink shows fluidic nature like solution at a low temperature (below 15 °C), but when it is in a high temperature similar to that of the human body (~37 °C), it is crosslinked spontaneously to increase its stiffness greatly. Pati et al. assessed thermal crosslinking of heart, cartilage, and adipose dECM bioinks, and those bioinks showed desirable sol–gel transitions from 15 to 37 °C (Fig. 6.2a) [4]. Moreover, they analyzed steady shear sweep to confirm shear thinning in a certain range of shear rate for extrusion and assessed storage (G') and loss (G'') moduli to verify temperature-dependent crosslinking and stiffnesses in each point (Fig. 6.2b). Rathan et al., after they checked the shear thinning behavior, analyzed thixotropic property of alginate/cartilage dECM bioink by testing recovery of viscosity, and they found that viscosity of the bioink was recovered immediately

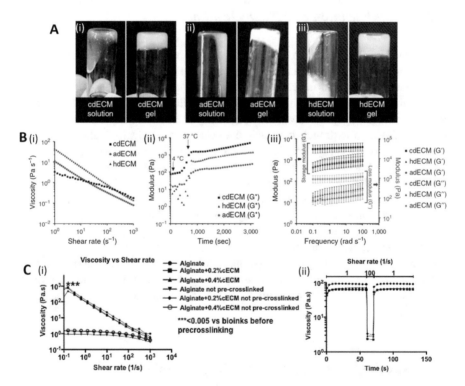

Fig. 6.2 Rheological property of dECM bioink. (**a**) Gelation of (i) cartilage dECM (cdECM), (ii) adipose dECM (adECM), and (iii) heart dECM (hdECM) bioink. (Reproduced with permission from [4]). (**b**) (i) Viscosity, (ii) gelation kinetics, and (iii) dynamic modulus of cdECM, adECM, and hdECM bioinks. (**c**) (i) Viscosity and (ii) thixotropic property of cartilage dECM bioink with or without mixing of alginate. (Copyright (2019) Wiley. Used with permission from [28])

in the shear rate range of 1–100 (s^{-1}) (Fig. 6.2c) [28]. Though these rheological properties are also very important, the actual value of each case cannot be overlooked. This is because the mechanical properties of each tissue are different from each other for their optimized functions, and the more similar they are, the more native-like tissue constructs can be fabricated [29]. Lee at al. controlled concentration of liver dECM bioink from 1.5% to 3.0% (w/v), and the viscosity and modulus showed approximately twofold increase [18]. Rathan et al. also prepared their alginate/cartilage dECM bioink in different concentrations from 0.2% to 0.4%, showing difference in the values [28].

6.4 3D Printing of Tissue/Organ Constructs Using dECM Bioinks

As discussed above, dECM bioinks contain abundant ECM components that directly affect cell viability, and there are also tissue-specific factors inherited from tissue origins. Therefore, the functionality of dECM bioink can be expected to show favorable performance in both cases. Pati et al. developed heart, cartilage, and adipose dECM bioinks and fabricated 3D cell constructs of each tissue analogue (Fig. 6.3a–c) [4]. The 3D cell-printed constructs using each dECM bioink maintained their pre-designed structure after crosslinking, and tissue-specific functions were dramatically improved compared to the corresponding tissue analogues using collagen bioink after 14 days culture. That study was the first to apply dECM materials to bioinks demonstrating the functionality and applicability of dECM bioink and has had a significant impact on other tissue engineering studies based on 3D printing. On the same principle, Kim et al. prepared pancreas dECM (pdECM) bioink to develop a 3D islet construct (Fig. 6.3d) [30]. The islet-printed construct showed similar viability in the case of using type I collagen bioink and pdECM bioink, but the latter responded more in vivo, likely due to high glucose condition by secreting more insulin than the former. For corneal reconstruction, Kim et al. developed a corneal dECM (Co-dECM) bioink, and the transparency, which is an essential property of cornea, was more similar to native cornea than that with type I collagen bioink (Fig. 6.3e) [31]. In the stem cell-derived keratocyte constructs, expression of corneal functional marker was higher in the Co-dECM group than type I collagen group. Altogether, dECM bioinks showed suitable mechanical properties for 3D cell printing technology, and their functionality was validated by improvement in tissue-specific behaviors compared to with conventional bioinks.

6.5 Variants of dECM Bioinks

Since dECM bioink follows rheological and mechanical properties of collagen, it does not show very suitable stiffness and takes a long time for crosslinking, implementing difficult-to-fabricate bulky structures. Moreover, there has been report that

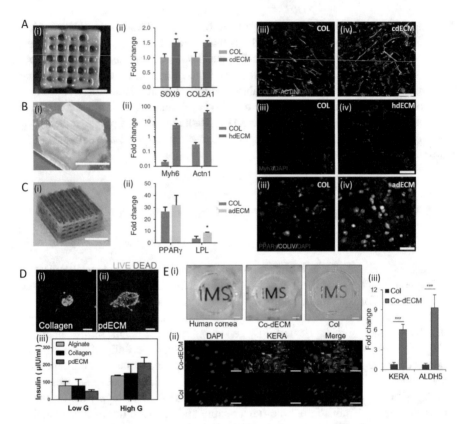

Fig. 6.3 Functional validation of 3D-printed cell constructs using dECM bioink. (**a**) 3D cell-printed cartilage tissue using cartilage dECM (cdECM) bioink and PCL framework (i) and upregulated chondrogenic gene (ii) and protein (iii, iv) expression compared to printed cell construct using type I collagen bioink. (Reproduced with permission from [4]). (**b**) 3D-printed cardiac structure using heart dECM (hdECM) bioink (i) and improved expression of cardiac functional markers in gene (ii) and protein (iii, iv) levels. (**c**) 3D adipose construct printed using adipose dECM (adECM) bioink (i), and upregulation of adipose-related gene (ii) and protein marker (iii, iv) expression. 3D cell-printed construct using type I collagen bioink was control group for (**a–c**). (**d**) 3D printing of islet construct using pancreas dECM (pdECM) bioink and viability of islet in the case of using (i) collagen bioink and (ii) pdECM bioink. (Reproduced from [30] with permission from The Royal Society of Chemistry). (iii) Improved insulin secretion in islet construct using pdECM bioink in high glucose condition. (**e**) Comparative transparency study of native human cornea, corneal dECM (Co-dECM) bioink, and type I collagen (Col) bioink (i), with higher expression of corneal functional protein (ii) and gene (iii) in Co-dECM group than Col group. (Reproduced from [31] with permission from SAGE). SOX9: SRY-box transcription factor 9; COL: collagen; Myh: myosin heavy chain; Actn: alpha-sarcomeric actinin; PPARγ: peroxisome proliferator-activated receptor gamma; LPL: low-density lipoprotein; Low G: low-glucose media; High G: high-glucose media; KERA: keratocan; ALDH: aldehyde dehydrogenase

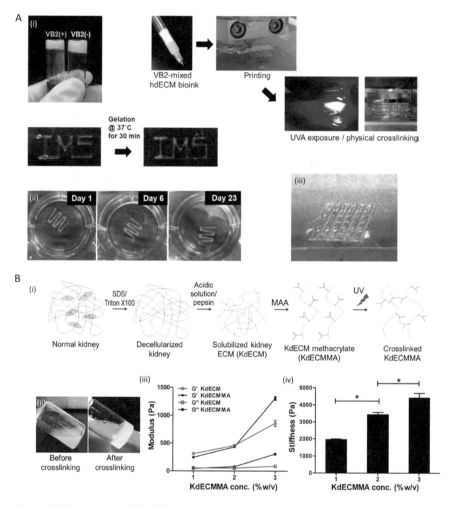

Fig. 6.4 Variants of dECM bioink to improve mechanical property. (**a**) (i) Addition of riboflavin (vitamin B2, VB2) to heart dECM (hdECM) bioink and 3D cell printing using it. (ii) Structural integrity of two-step crosslinked constructs using hdECM bioink with VB2 until day 23 and (iii) 10-layer printed cell constructs. (Reprinted from [32], Copyright (2016), with permission from Elsevier). (**b**) (i) Methacrylation of kidney dECM (KdECM). (ii) Crosslinking of KdECM methacrylate (KdECMMA) and improvement of (iii) modulus and (iv) stiffness by the chemical modification. (Copyright (2019) Wiley. Used with permission from [33]). G' storage modulus, G'' loss modulus

the dECM bioink showed low stiffness compared to the target tissue, even after its thermal crosslinking. To address this issue, it has been reported that dECM bioink is added with a crosslinker or mixed with other biomaterials and chemically modified to improve it. Jang et al. put riboflavin (vitamin B2, VB2) to heart dECM bioink to improve mechanical property by secondary photo-crosslinking (Fig. 6.4a) [32].

The mixing of 0.2% (w/v) VB2 into the hdECM bioink was similar to the viscosity before addition. After primary thermal crosslinking, secondary photo-crosslinking was carried out with UVA, and it resulted in increasing of stiffness about 33-fold, which is similar to native cardiac tissue. The printed structure was maintained until day 23, and it was available to fabricate ten-layered constructs without collapsing the structure. Moreover, when cardiac progenitor cell was printed with the advanced hdECM bioink, heart-specific genes such as GATA4, Nk2 homeobox (Nkx 2.5), and cardiac Troponin I (cTnI) were upregulated significantly. Ali et al. chemically modified kidney dECM (KdECM) bioink by methacrylation to make printability better (Fig. 6.4b) [33]. After modification, methacrylated KdECM bioink (KdECMMA) showed improved modulus and stiffness after crosslinking. Especially, 3% (w/v) of KdECMMA had suitable shape fidelity to fabricate bulky constructs, and viability of renal cells also improved in the KdECMMA-treated group compared to the naïve one. Moreover, enzymatic activity such as hydrolase and gamma-glutamyltranspeptidase of renal cell-laden constructs using KdECMMA remained high until 2 weeks culture.

References

1. Gilbert TW, Sellaro TL, Badylak SF. Decellularization of tissues and organs. Biomaterials. 2006;27(19):3675–83.
2. Hoganson DM, O'Doherty EM, Owens GE, Harilal DO, Goldman SM, Bowley CM, Neville CM, Kronengold RT, Vacanti JP. The retention of extracellular matrix proteins and angiogenic and mitogenic cytokines in a decellularized porcine dermis. Biomaterials. 2010;31(26):6730–7.
3. Elliott RA Jr, Hoehn JG. Use of commercial porcine skin for wound dressings. Plast Reconstr Surg. 1973;52(4):401–5.
4. Pati F, Jang J, Ha D-H, Kim SW, Rhie J-W, Shim J-H, Kim D-H, Cho D-W. Printing three-dimensional tissue analogues with decellularized extracellular matrix bioink. Nat Commun. 2014;5:3935.
5. Choudhury D, Tun HW, Wang T, Naing MW. Organ-derived decellularized extracellular matrix: a game changer for bioink manufacturing? Trends Biotechnol. 2018;36(8):787–805.
6. Denner J, Tönjes RR. Infection barriers to successful xenotransplantation focusing on porcine endogenous retroviruses. Clin Microbiol Rev. 2012;25(2):318–43.
7. Moses HL, Roberts AB, Derynck R. The discovery and early days of TGF-β: a historical perspective. Cold Spring Harb Perspect Biol. 2016;8(7):a021865.
8. Habiro K, Sykes M, Yang YG. Induction of human T-cell tolerance to pig xenoantigens via thymus transplantation in mice with an established human immune system. Am J Transplant. 2009;9(6):1324–9.
9. Komoda H, Okura H, Lee CM, Sougawa N, Iwayama T, Hashikawa T, Saga A, Yamamoto-Kakuta A, Ichinose A, Murakami S. Reduction of N-glycolylneuraminic acid xenoantigen on human adipose tissue-derived stromal cells/mesenchymal stem cells leads to safer and more useful cell sources for various stem cell therapies. Tissue Eng Part A. 2009;16(4):1143–55.
10. Sandrin M, Mckenzie IF. Galα (1, 3) Gal, the major xenoantigen (s) recognised in pigs by human natural antibodies. Immunol Rev. 1994;141(1):169–90.
11. Crapo PM, Gilbert TW, Badylak SF. An overview of tissue and whole organ decellularization processes. Biomaterials. 2011;32(12):3233–43.
12. Kasimir M-T, Rieder E, Seebacher G, Silberhumer G, Wolner E, Weigel G, Simon P. Comparison of different decellularization procedures of porcine heart valves. Int J Artif Organs. 2003;26(5):421–7.

13. Gallagher SR. Quantitation of DNA and RNA with absorption and fluorescence spectroscopy. Curr Protoc Neurosci. 2011;56(1):A.1K.1–A.1K.14.
14. Noothi SK, Kombrabail M, Kundu TK, Krishnamoorthy G, Rao BJ. Enhanced DNA dynamics due to cationic reagents, topological states of dsDNA and high mobility group box 1 as probed by PicoGreen. FEBS J. 2009;276(2):541–51.
15. Barbosa I, Garcia S, Barbier-Chassefière V, Caruelle J-P, Martelly I, Papy-García D. Improved and simple micro assay for sulfated glycosaminoglycans quantification in biological extracts and its use in skin and muscle tissue studies. Glycobiology. 2003;13(9):647–53.
16. Hofman K, Hall B, Cleaver H, Marshall S. High-throughput quantification of hydroxyproline for determination of collagen. Anal Biochem. 2011;417(2):289–91.
17. Kim BS, Kwon YW, Kong J-S, Park GT, Gao G, Han W, Kim M-B, Lee H, Kim JH, Cho D-W. 3D cell printing of in vitro stabilized skin model and in vivo pre-vascularized skin patch using tissue-specific extracellular matrix bioink: a step towards advanced skin tissue engineering. Biomaterials. 2018;168:38–53.
18. Lee H, Han W, Kim H, Ha D-H, Jang J, Kim BS, Cho D-W. Development of liver decellularized extracellular matrix bioink for three-dimensional cell printing-based liver tissue engineering. Biomacromolecules. 2017;18(4):1229–37.
19. Welham NV, Chang Z, Smith LM, Frey BL. Proteomic analysis of a decellularized human vocal fold mucosa scaffold using 2D electrophoresis and high-resolution mass spectrometry. Biomaterials. 2013;34(3):669–76.
20. Wiles K, Fishman JM, De Coppi P, Birchall MA. The host immune response to tissue-engineered organs: current problems and future directions. Tissue Eng Part B Rev. 2016;22(3):208–19.
21. Fujii K, Yamagishi T, Nagafuchi T, Tsuji M, Kuboki Y. Biochemical properties of collagen from ligaments and periarticular tendons of the human knee. Knee Surg Sports Traumatol Arthrosc. 1994;2(4):229–33.
22. Fratzl P. Collagen: structure and mechanics, an introduction. Collagen. New York, NY: Springer; 2008. p. 1–13.
23. Samuel CS. Determination of collagen content, concentration, and sub-types in kidney tissue. Kidney research. New York, NY: Springer; 2009. p. 223–35.
24. Nalinanon S, Benjakul S, Visessanguan W, Kishimura H. Tuna pepsin: characteristics and its use for collagen extraction from the skin of threadfin bream (Nemipterus spp.). J Food Sci. 2008;73(5):C413–9.
25. Li H, Tan YJ, Leong KF, Li L. 3D bioprinting of highly thixotropic alginate/methylcellulose hydrogel with strong interface bonding. ACS Appl Mater Interfaces. 2017;9(23):20086–97.
26. Liu W, Heinrich MA, Zhou Y, Akpek A, Hu N, Liu X, Guan X, Zhong Z, Jin X, Khademhosseini A. Extrusion bioprinting of shear-thinning gelatin methacryloyl bioinks. Adv Healthc Mater. 2017;6(12):1601451.
27. Hölzl K, Lin S, Tytgat L, Van Vlierberghe S, Gu L, Ovsianikov A. Bioink properties before, during and after 3D bioprinting. Biofabrication. 2016;8(3):032002.
28. Rathan S, Dejob L, Schipani R, Haffner B, Möbius ME, Kelly DJ. Fiber reinforced cartilage ECM Functionalized bioinks for functional cartilage tissue engineering. Adv Healthc Mater. 2019;8:e1801501.
29. Cox TR, Erler JT. Remodeling and homeostasis of the extracellular matrix: implications for fibrotic diseases and cancer. Dis Model Mech. 2011;4(2):165–78.
30. Kim J, Shim IK, Hwang DG, Lee YN, Kim M, Kim H, Kim S-W, Lee S, Kim SC, Cho D-W. 3D cell printing of islet-laden pancreatic tissue-derived extracellular matrix bioink constructs for enhancing pancreatic functions. J Mater Chem B. 2019;7(10):1773–81.
31. Kim H, Park M-N, Kim J, Jang J, Kim H-K, Cho D-W. Characterization of cornea-specific bioink: high transparency, improved in vivo safety. J Tiss Eng. 2019;10:2041731418823382.
32. Jang J, Kim TG, Kim BS, Kim S-W, Kwon S-M, Cho D-W. Tailoring mechanical properties of decellularized extracellular matrix bioink by vitamin B2-induced photo-crosslinking. Acta Biomater. 2016;33:88–95.
33. Ali M, Yoo JJ, Zahran F, Atala A, Lee SJ. A photo-crosslinkable kidney ECM-derived bioink accelerates renal tissue formation. Adv Healthc Mater. 2019;8:e1800992.

Chapter 7
Various Applications of 3D-Bioprinted Tissues/Organs Using Tissue-Specific Bioinks

7.1 Skin

7.1.1 Structural and Physiological Features of the Skin

Human skin is the largest organ, with an average thickness of 2.5 mm. This tissue provides a barrier function and protects against harmful UV rays, physical damage, and water loss [1]. Human skin can be categorized into two different regions (upper epidermal and lower dermal), which are separated by a basement membrane layer positioned at the epidermal–dermal junction (Fig. 7.1).

The upper epidermal region is characterized by densely packed keratinocytes that form stratified cell layers with increasing degree of differentiation near the skin surface. The proliferating keratinocytes residing at the epidermal–dermal junction undergo sequential differentiation to form fully stratified epidermal layers over a period of 2 weeks. This stratified epidermis provides the critical skin barrier function [3]. The lower dermal region is characterized by the presence of collagen fibers with relatively low fibroblast density (0.2–2.0×10^5 fibroblasts/cm^3). The dermal compartment can be again classified into upper papillary and lower reticular regions. The upper papillary dermal region comprises densely packed collagen fibers (higher ratio of type III to type I collagen) that are randomly oriented, whereas the lower reticular dermal region comprises porous and thick collagen fibers (higher ratio of type I to type III collagen). This 3D hierarchical collagen-based microenvironment regulates the proliferation of fibroblasts and their ECM secretion. The dermal fibroblasts are responsible for the secretion of important ECM proteins (collagen I, IV, laminin, fibronectin, and elastin) and growth factors, which play a key role in promoting cell–ECM and cell–cell interactions. Notably, supplementing the culture medium with biomolecules results in significant improvements in ECM deposition by fibroblasts [4]. The basement membrane at the epidermal–dermal junction is a macromolecular structure with highly complex ultrastructural and biochemical cues, and it is critical for positional orientation of the melanocytes. The basement

© Springer Nature Switzerland AG 2019
D.-W. Cho et al., *3D Bioprinting*, https://doi.org/10.1007/978-3-030-32222-9_7

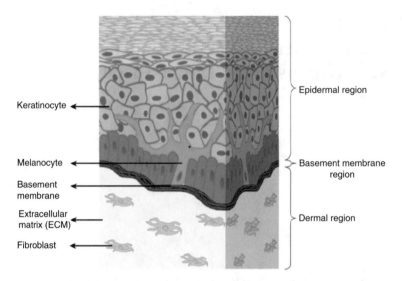

Fig. 7.1 Schematic drawings of native human skin anatomy. (Reproduced with permission from [2])

membrane layer, which is composed of collagen IV, VII, and laminin, is mainly produced by keratinocyte-fibroblast interactions in an organotypic co-culture. The melanocytes representing skin color are located on the dermal region together with basal keratinocytes. The ratio of melanocytes to the basal keratinocytes at the epidermal–dermal junction is approximately 1:20, and a minimum melanocyte density of 1.0×10^4 cells/cm^2 is required to present the skin pigmentation in tissue-engineered skin constructs. The melanocytes synthesize the melanin pigment, which provides the distinctive skin color in each individual and offers protection against UV radiation. The synthesized melanin pigments are stored within melanosomes, which are transferred to overlaying suprabasal keratinocytes via elongated dendrites. This melanosome uptake is regulated by the degree of keratinocytes differentiation.

7.1.2   3D Bioprinting of In Vitro Skin Model

Given the purpose of precise cosmetics and drug development, there is a clear need for engineering in vitro skin models mimicking the complexity of native skin (Fig. 7.2) [5].

A key hope for 3D bioprinting of the skin is to realize the recapitulation of such complexity of skin anatomy through precise deposition of various cells and biomaterials [2]. Related to 3D bioprinting of the skin, a pioneering study demonstrated the potential of inkjet 3D bioprinting for engineering skin tissue by using keratinocytes and fibroblasts as constituent cells (Fig. 7.3a) [6, 7]. Histology and

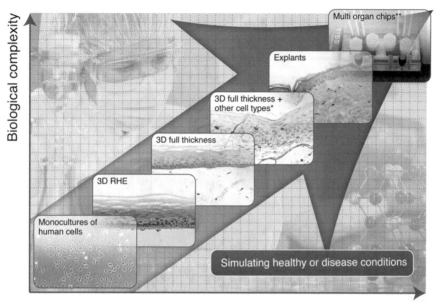

Fig. 7.2 Increasing need for engineered in vitro skin models with higher biological complexity. (Reproduced with permission from [5])

immunofluorescence characterization were evaluated. Although this study showed the potential of 3D bioprinting to engineer skin tissue, the 3D-bioprinted skin was morphologically and biologically far from native human skin. In another study, a laser-assisted 3D bioprinting was used to recreate a skin substitute by positioning fibroblasts and keratinocytes on top of a stabilizing matrix (Matriderm) [8]. Although two types of cells were precisely positioned on the predefined location, the printed constructs did not undergo air–liquid interface, which is essential for epidermal stratification (Fig. 7.3b). For this reason, the constructs could not be fully differentiated and matured. Indeed, the histological images were considerably different from those of native human skin.

Undoubtedly, type I collagen, used in the aforementioned studies, is the ECM composition representing native human dermis. However, its weak mechanical properties result in significant contraction, ultimately limiting their use in elucidating the mechanisms regulating balanced growth and differentiation in a stabilized native tissue. To overcome the shrinkage of type I collagen, a study formulated a mixture of 10% (w/v) bovine gelatin, 0.5% (w/v) low viscosity alginate, and 2% (w/v) fibrinogen [10]. The formulated bioink enabled the fabrication of centimeter-scale complex structures filling features of 200 μm width (Fig. 7.4a). The 3D-bioprinted skin using this mixture was stabilized for 26 days of culture. The morphology of matured skin model was remarkably similar to that of native human skin (Fig. 7.4b). Although this mixture consisting of alginate, fibrinogen, and gelatin

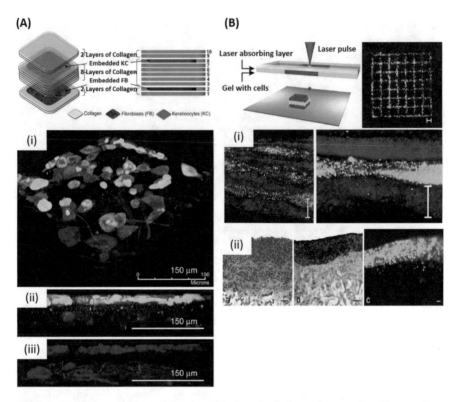

Fig. 7.3 (**a**) 3D normal skin models using inkjet-based printing; volume-rendered immunofluorescent images of multilayered printing of keratinocytes and fibroblasts (i) and its projection of (ii) keratin-containing epidermis layers and (iii) β-tubulin-containing keratinocytes and fibroblasts. (**b**) 3D normal skin model using laser-assisted bioprinting technique; (i) a skin construct of fibroblasts (green) and keratinocytes (red) was printed, and hematoxylin and eosin (H&E) staining of bioprinted cells was investigated in a skin tissue-like pattern (ii). (Reproduced with permission from [6–9])

could act the part in a stabilized substrate during tissue maturation, each constituent of the mixed bioink is not the ECM representing native dermal ECM. ECMs play a crucial role in enhancing cell–cell and cell–ECM interactions, not merely holding cells and forming the architecture [11]. Through such interactions, cells produce biological and biochemical cues and lead to tissue development toward native-like microenvironment. This means that embedded skin cells within such artificial mixture are unable to physiologically receive favorable cues, including cytokines and growth factors. Eventually, this might cause unexpected difficulty in accurately predicting cosmetics and drugs under development.

To replicate a more realistic in vivo microenvironment, skin-derived decellularized extracellular matrix (S-dECM) was considered a promising source as a bioink [12]. This bioink was used to print in vitro dermal/epidermal skin models. Unlike the collagen group, dECM-based skin model was stabilized during in vitro tissue

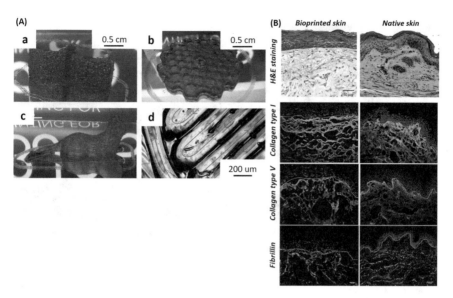

Fig. 7.4 (**a**) Various-shaped 3D-printed constructs using the alginate–fibrinogen–gelatin bioink: (a) a water-tight structure with two compartments filled with blue-dyed liquid; (b) a complex structure with honeycomb filling features of 200 μm width; (c) a centimeter-scale complex object with overhanging structures; (d) an enlarged view of printed lines. (**b**) Immunofluorescent observations of epidermal differentiation and dermal markers' profiles of bioprinted skin in comparison to native human skin from healthy donor. (Reproduced with permission from [10])

culture, possibly owing to thicker collagen fibers and remaining other ECM components such as elastin and hyaluronic acid (Fig. 7.5a). This stabilized substrate allowed for better organization of epidermal differentiation (Fig. 7.5b). Furthermore, this bioink was used to engineer a 3D prevascularized skin patch that can promote wound healing in vivo. The results revealed that the dECM-based patch accelerated wound closure, reepithelization, and neovascularization as well as recovery of blood flow. For a further analysis, growth factor and cytokine array analysis were conducted. The resultant data showed the dECM bioink contains more various growth factors and cytokines when compared with type I collagen; in particular, wound healing-related factors (EGF and bFGF) in the dECM bioink were significantly higher than those in the collagen. These results support that S-dECM is a promising bioink source for advanced skin tissue engineering.

Most recently, based on S-dECM bioink, a 3D-bioprinted perfusable and vascularized skin model composed of the epidermis, dermis, and hypodermis was engineered for the first time [13, 14]. To incorporate the perfusable channel between the dermal and hypodermal compartments, polycaprolactone-based transwell platform was suggested, and gelatin-based vascular bioink was used as a sacrificial structure; all fabrication processes are shown in Fig. 7.6a. The perfusion ability was confirmed by flowing red ink through inlet and outlet of the channel (Fig. 7.6b). The engineered advanced skin model had more similarities to native human skin compared with the dermal and epidermal skin model, indicating that it better reflects the actual

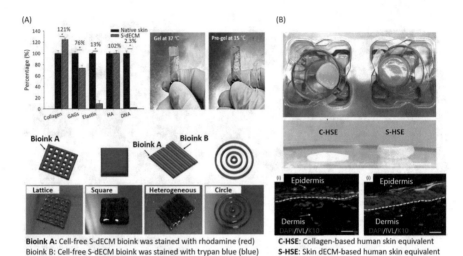

Bioink A: Cell-free S-dECM bioink was stained with rhodamine (red)   C-HSE: Collagen-based human skin equivalent
Bioink B: Cell-free S-dECM bioink was stained with trypan blue (blue)   S-HSE: Skin dECM-based human skin equivalent

Fig. 7.5 Development of skin-derived decellularization extracellular matrix bioink; (**a**) analysis of remaining ECM composition after decellularization, confirmation of thermal gelation, and test for printability. (**b**) Engineered in vitro skin model by using either type I collagen or the developed dECM bioink; immunofluorescent images of epidermal differentiation markers in C-HSE (i) and S-HSE (ii) (involucrin (IVL), late differentiation marker; K10, early differentiation marker of epidermis). (Reproduced with permission from [12])

complexity of native human skin (Fig. 7.6c). Although the ultimate goal of 3D bioprinting of the skin with complete functional performances has yet to be achieved, 3D bioprinting of the skin definitely shows promise in several critical aspects of skin tissue engineering, including creating pigmented and/or aging skin models and hair follicles.

7.2   Blood Vessels

7.2.1   Structural and Physiological Functions of Blood Vessels

As a part of the circulatory system, the vascular system serves to transport blood to every corner of the human body. Pumped out from the heart, the fresh blood provides abundant nutrients and oxygen to cells to support their survival and functions. In the latter half of the circulatory journey, the blood is depleted of vital supports but carries the waste and carbon dioxides removed from tissues back to the heart. Undoubtedly, the blood vessel system is one of the most important to sustain healthy life, as all of the body's tissues rely on its functionalities.

Although a blood vessel generally shows as a simple tube, it can be categorized into a variety of types that present distinct structures and compositions (Fig. 7.7).

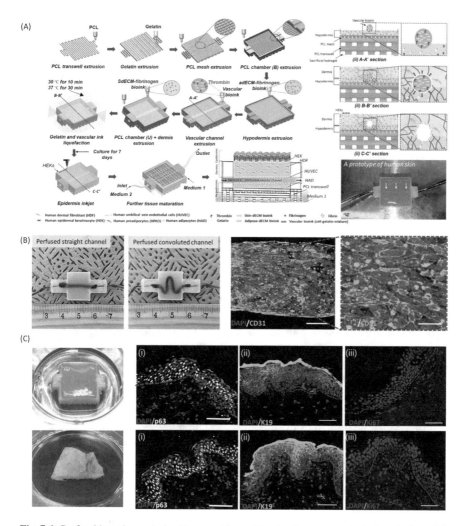

Fig. 7.6 Perfusable and vascularized human skin model using the suggested printing platform. (**a**) The whole processes for engineering perfusable skin model composed of epidermis, dermis, and hypodermis. (**b**) The confirmation of the ability to perfuse the red ink into the channel. (**c**) Histological observation of the developed skin model (upper) and native human skin (lower); p63 (i) and K19 (ii) denote stemness markers of epidermis, and Ki67 (iii) represents proliferative state of the skin. (Reproduced with permission from [13])

Structurally, the vascular system resembles a tree that consists of main stems and descending branches. In particular, the aorta, directly originating from the heart ventricle, is the biggest blood vessel (~25 mm inner diameter (ID)) that receives and carries the fresh blood as much as possible. Branching from the aorta, the arteries and arterioles become smaller and smaller (artery, 1–4 mm ID; arteriole, 30–100 μm ID), splitting the stream toward different directions throughout the human body. In approaching tissues and organs, the blood flow slows down and enters the narrowest

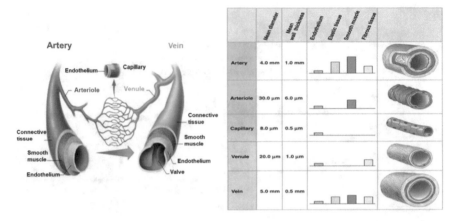

Fig. 7.7 Schematic diagram of blood vessel system in the human body and categorization depending on dimensions, structures, and compositions. (Copyright © 2009 Pearson Education)

vessels, named capillaries (<10 μm ID), in which the exchanging of nutrients, oxygen, and metabolites between tissues and blood takes place [15]. Afterward, the nutrient-depleted blood is transported through the venules, converging to veins, which eventually return to the heart. In addition to the structural disparity, the composition also varies in each type of blood vessel (Fig. 7.7). Typically, blood vessels are composed of three concentric layers in anatomy, which are the tunica intima (in the internal site), tunica media (in middle), and tunica externa (as the external shell). Each layer contains specific cells and extracellular matrix (ECM) components that determine their distinct roles in blood transportation.

All of the constituents of blood vessels contribute differently to regulating the vascular functions. The tunica intima includes endothelium and an internal elastic membrane. The endothelium is a single layer composed of endothelial cells (ECs), which is of the great importance of mitigating blood coagulation and regulating the transportation of metabolic solutes (e.g., water, ions, and molecules). The endothelium is comprised of over 60 trillion cells from 100,000 km of interconnected vessels with a surface area of 4000 m² [16]. When facing external stimulations, such as inflammation and hypoxia, ECs can respond to support the trafficking of immune cells from blood to tissues and to generate neovasculature through angiogenic sprouting. Beneath the endothelium, a thin elastic membrane separates the tunica intima and tunica media to prevent migration of smooth muscle cells (SMCs). Tunica media is made up of SMCs and elastic protein fibers composed of collagen and elastin. The circumferentially arranged SMCs play a critical role in adapting the blood pressure by contracting or relaxing, whereas elastic fibers allow vessels to stretch and recoil [17]. The tunica externa, which contains loose collagen fibers secreted by fibroblasts, exists to protect blood vessel, improve tensile stress, and anchor the structure to adjacent tissues. In addition, besides the dominated fibroblasts, the external layer also includes a mixture of progenitor/stem cells, pericytes, myofibroblasts, macrophages, and dendritic cells, which respond to endothelial injury and travel into the two inner layers for the corrective mission of ECM composition [18].

7.2.2   3D Bioprinting of In Vitro Vascular Model

The dysfunctions of the vascular system are closely related to some critical diseases such as tumor angiogenesis, cancer metastasis, tissue edema, inflammation, and diseases of blood vessel themselves. Therefore, there has been enduring interest in the study of physiological functions and pathological changes of blood vessels, as well as the relevant diseases. Previous efforts have developed a variety of fabrication strategies to create the in vitro vascular models.

3D bioprinting has been used to construct hydrogel built-in structures as the in vitro vascular models. One strategy is to print sacrificial molds. This method usually 3D bioprints a template of the targeting vascular structures using fugitive biomaterials (e.g., pluronic F127 [19–21], gelatin [22], agarose [23], and carbohydrate glass [24]). After casting and gelation of ECM hydrogels, the printed templates can be removed by either triggering dissolution of the fugitive biomaterials or aspirating them manually, leaving the interconnected hollow channels in the hydrogel matrix. Upon seeding of ECs in these microchannels, the adhered cells generate an endothelium layer on the luminal surface of the channel, revealed by staining against VE-cadherin, an endothelial adhesion junction marker. Using this approach, vascularized thick tissues (>1 cm thickness) have been successfully fabricated [20] (Fig. 7.8). Due to the presence of embedded vasculatures, it progressively overcomes the limited survival of cells in thick in vitro tissue models. When involving human mesenchymal stem cells (hMSCs) and human neonatal dermal fibroblasts in the hydrogel matrix, the perfusable vasculatures allow for the supply of growth factors to differentiate hMSCs toward an osteogenic lineage in situ.

Similar to the sacrificial 3D bioprinting, another indirect approach prints the removable materials directly into the supporting matrix to template the vasculatures. The maintenance of printed structures requires the materials of the matrix to exhibit a Bingham plastic property, which does not yield unless a threshold shear force is reached. So far, a variety of biomaterials with such characteristics have been explored, such as gelatin granule, Carbomer, and chemically modified hyaluronic acid (HA) [25–27]. The advantage of this method over the sacrificial 3D bioprinting is the ability to fabricate the vasculature in the Z-axis direction. In a recent report, adamantine HA (ad-HA) modified with norbornene (AdNor-HA) was used as fugitive material and matrix material, respectively, to 3D bioprint the microchannels embedded in hydrogel [28]. This technique permits the printing of even stenosis and spiral channels (Fig. 7.9). Using a model containing two adjacent channels to provide a gradient growth factor to the endothelium, the researchers studied the influence of vascular geometries on the progress of directional angiogenesis.

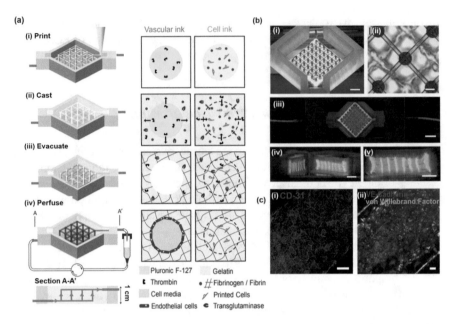

Fig. 7.8 Three-dimensional vascularized tissue fabrication. (**a**) Schematic illustration of the fabrication process. (i) Fugitive (vascular) ink, which mixes pluronic and thrombin, and cell-laden inks, which contain gelatin, fibrinogen, and cells, are printed within a 3D perfusion chip. (ii) ECM material, which combines gelatin, fibrinogen, cells, thrombin, and transglutaminase (TG), is then cast over the printed inks. After casting, thrombin induces fibrinogen cleavage and rapid polymerization into fibrin both in the cast matrix, and through diffusion, in the printed cell ink. Similarly, TG diffuses from the molten casting matrix and slowly crosslinks the gelatin and fibrin. (iii) Upon cooling, the fugitive ink liquefies and is evacuated, leaving behind a pervasive vascular network, which is (iv) endothelialized and perfused via an external pump. (**b**) (i) Photographs of interpenetrated sacrificial (red) and cell inks (green) as printed on chip (scale bar: 2 mm), (ii) a top-down bright-field image of sacrificial and cell inks (scale bar: 50 μm), (iii) photograph of a printed tissue construct housed within a perfusion chamber, and (iv, v) corresponding cross sections (scale bars: 5 mm). (**c**) Confocal microscopy image of the vascular network after 42 days, CD-31 (red), vWF (blue), and VE-cadherin (magenta) (scale bars: 100 μm). (Reproduced with permission from [20])

Despite the indirect 3D bioprinting being able to fabricate the microchannels in the hydrogel, the seeding of endothelial cells into the tubular channels is a somewhat unstable method to form endothelium. The cell seeding efficiency is highly dependent on numerous factors, including the cellular affinity of the matrix material, the roughness of channel surface, seeding cell density, and seeding time. In addition, inverting the models is usually required to seed the cells on the top of the tubular channel.

3D bioprinting is also capable of organizing cell strands and aggregates to fabricate the vascular constructs (i.e., scaffold-free 3D bioprinting). Upon the culture, the patterned individual cell pellets can be fused together to generate intact vessel-like structures. Since the cell aggregates cannot hold their printed positions, especially in the vertical direction, it is indispensable to use additional assisting tools to

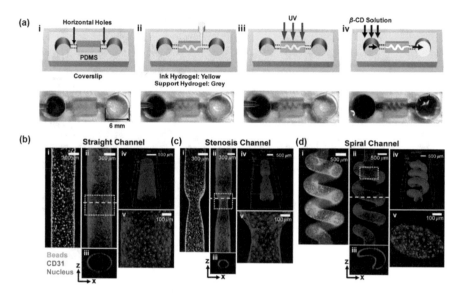

Fig. 7.9 Microchannel fabrication process through the 3D bioprinting of a fugitive ink hydrogel into a support hydrogel. (**a**) Schematic of the microchannel fabrication process where (i) a PDMS holder is placed on a coverslip, (ii) the PDMS reservoir is filled with a support hydrogel (gray) and an ink hydrogel (yellow) is printed within, (iii) the hydrogel is exposed to light in the presence of a photoinitiator to stabilize the support hydrogel, and (iv) the fugitive ink hydrogel is washed with flow and the introduction of excess β-cyclodextrin (β-CD) in solution. Fluorescent images showing the endothelial cell-seeded channels with various designs, including (**b**) a straight channel, (**c**) a stenosis channel, and (**d**) a spiral channel. (Reproduced with permission from [28])

retain the fabricated structures until the fusion process is completed. One representative example deposited cylindrical agarose to build a mold before printing the cell spheroids [29]. In a layer-by-layer manner, the cell spheroids were embedded in the tubular mold made from agarose filaments. As the spheroids fused together and matured toward vascular tissue, the mold was removed, leaving behind the tissue (Fig. 7.10a). Using this method, multiple-layered and bifurcated vessels (OD ranging from 0.9 to 2.5 mm) containing human SMCs and human fibroblasts were successfully constructed.

In addition to the use of the sacrificial mold, a more sophisticated 3D bioprinter was developed to pinpoint multicellular spheroids into a needle-array platform. This bioprinter can aspirate a single-cell spheroid generated from multiple cell types (ECs, SMCs, and fibroblasts) by a robotically controlled fine suction nozzle (OD of 0.45 mm and ID of 0.25 mm) from a 96-well plate and place it into the needle array [30]. Using this method, a tubular structure with 1.5 mm diameter and 7 mm length was bioprinted by plotting a total of 500 cell spheroids into a 9 × 9 needle array within 1.3 h (Fig. 7.10b). After 4 days of culture, the printed spheroids fused to form tubular tissue, allowing for the removal of the needle array. By remodeling the tissue using a designed perfusion platform for 4 days, abundant collagen was synthesized, probably from fibroblasts. On the other hand, after the implantation into nude

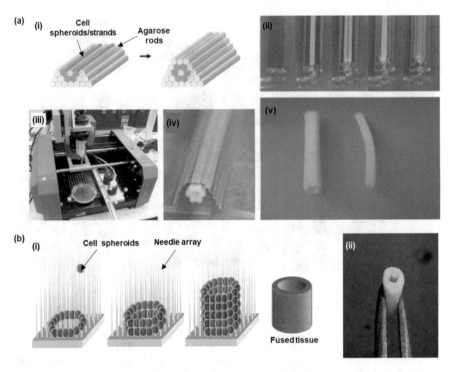

Fig. 7.10 3D bioprinting of scaffold-free vascular tissues. (**a**) (i) Schematic diagram of 3D bio-printing of tubular structures using cellular strands with the assistance of sacrificial agarose rods; (ii) layer-by-layer deposition of agarose cylinders and multicellular SMC strands; (iii) the bio-printer outfitted with two vertically moving print heads. (iv) The printed construct; (v) the engi-neered SMC tubes of distinct diameters resulted after 3 days of post-printed fusion (left, 2.5 mm OD; right, 1.5 mm OD) [29]. (**b**) (i) Schematic of 3D bioprinting of vascular structures by organiz-ing cell spheroids into a needle array; (ii) a scaffold-free vascular tissue is generated from the fusion of cell spheroids. (Reproduced with permission from [30])

rats for 5 days, the endothelial cells were arranged to generate an endothelial layer in the luminal surface.

Although the scaffold-free strategies have been realized to construct the vascular equivalent using only cells, it requires sophisticated 3D bioprinters to accurately organize the cell spheroids for fabricating the tubular structures. Meanwhile, to gen-erate cell spheroids, additional operation procedures should be implemented, leading to a complicated construction process. Moreover, the cell spheres are densely integrated, which might cause hypoxia in the central region. Hence, consid-ering the large size and high cell density, it is critical to minimize the hypoxia to support cell survival.

To overcome the challenges of unreliable ECs seeding in hollow channels and the complex fabrication of positioning cell spheroids, a novel fabrication technique, namely, 3D coaxial bioprinting, was developed. This method is capable of directly constructing freestanding vessel-like structures. Based on a coaxial nozzle that is

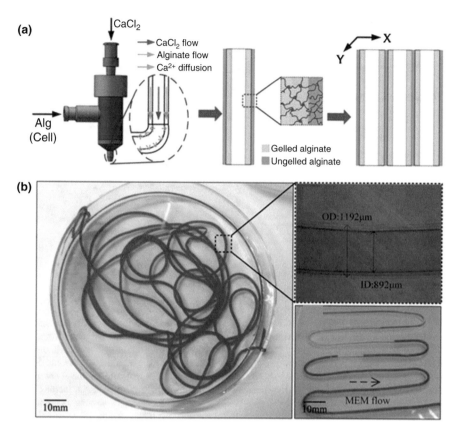

Fig. 7.11 3D coaxial bioprinting of freestanding vessel-like structure. (**a**) By co-extrusion of cell-laden alginate solution and calcium chloride (CaCl$_2$) from a coaxial nozzle, the alginate can be immediately crosslinked. (**b**) Continuous perfusable tubes can be fabricated using this technique. (Reproduced with permission from [31])

composed of a core and a shell needle, the 3D coaxial bioprinting technique can co-print two different materials simultaneously to produce a core/shell filament (Fig. 7.11). Upon the gelation of shell and removal of the core, a hollow conduit can be achieved. To avoid the collapse of the tubular structures, the shell materials need to be instantly crosslinked as extruded out.

Alginate has been widely used in this technique due to its immediate gelation upon exposure to calcium ions. Several efforts have successfully 3D-bioprinted perfusable vessels using alginate. During the printing process, cell-laden alginate and calcium choroid were dispensed through the shell and core needle, respectively, to crosslink the alginate bioink, forming a tubular structure [31, 32]. However, although many studies have reported the direct fabrication of perfusable tubes using the 3D coaxial bioprinting technique, functional vascular tissues have been realized because of the limited ability of alginate for promoting cell functionalities. Some other studies attempted mixing methacrylated gelatin, a widely used biomaterial that can

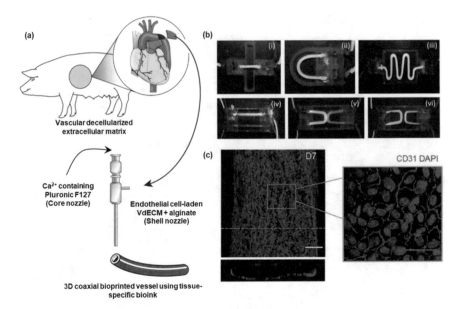

Fig. 7.12 3D coaxial bioprinting of vascular tissues using tissue-specific bioink. (**a**) The vascular tissue-specific bioink was derived from vascular decellularized extracellular matrix of porcine aortic tissue. After mixing with alginate (ratio 3:2), it supports the 3D bioprinting of vessel-like structure using a coaxial nozzle by co-dispensing the bioink and Ca^{2+} containing pluronic F127. (Reproduced with permission from [35]). (**b**) This tissue-specific bioink facilitates the 3D bioprinting technique to directly fabricate the freestanding conduits with various patterns, including (i) straight vessel, (ii) curve vessel, (iii) convoluted vessel, (iv) dual-parallel vessels, (v) attached dual-curve vessels, and (vi) discrete dual vessels. (**c**) Because the tissue-specific bioink provides a natural microenvironmental niche to compensate for the negative effect of alginate, the endothelial cells were encouraged to function and generate an intact endothelium tissue after 7 days (scale bar, 100 μm). (Reproduced with permission from [36])

support cell activity, to promote the functionality of embedded cells [33, 34]. However, a functional vascular tissue has not yet been achieved, though the cell proliferation was improved.

A vascular decellularized extracellular matrix (VdECM) originated from porcine aortic tissue was developed to compensate for the drawback of alginate [35] (Fig. 7.12a). After a series of decellularization and acidic digestion treatments, the aortic tissue can be formulated as printable bioink. This material inherits the complex ingredients of ECMs in natural tissue, including collagen, GAGs, and elastin. Hence, it offers a cell-favorable microenvironment to benefit cell activity. An optimized combination of the VdECM and alginate (3% VdECM and 2% alginate, 3V2A) was shown to be capable of enhancing the cell functionalities of ECs compared to type I collagen hydrogel. Moreover, this 3V2A hybrid bioink facilitates the direct 3D coaxial bioprinting of perfusable vessel-like structures with various predesigned patterns (Fig. 7.12b). Therefore, the human endothelial progenitor cells (HEPCs) or human umbilical vein endothelial cells (HUVECs) encapsulated in the printed tubes actively proliferate, migrate, and differentiate to form an endothelium tissue (Fig. 7.12c), which is difficult to achieve using other approaches. In a recent study, this printed endothelialized vessel was further applied to create in vitro vascular models that can recapitulate the physiological functions of endothelium, including the selective permeability, anti-platelet adhesion, shear-induced self-organization, and directional angiogenic sprouting [36]. In addition, when facing inflammatory stimulations, this vascular model exhibits pathological changes such as endothelium activation and ECs barrier breakdown.

Undoubtedly, the development of vascular tissue-specific bioink marked a breakthrough of materials deficiency and allowed to advance the 3D coaxial bioprinting for engineering in vitro vascular models. Although its current application only produces a single-layered conduit for recapitulating the endothelium pathophysiology, augmenting the numbers core/shell needles in the coaxial nozzle (e.g., triple- and quad-coaxial nozzle) could extend its potential to engineered multiple-layered tubes that are biomimetic to macrovessels such as arteries and veins.

7.3   Liver

7.3.1   Structural and Physiological Functions of the Liver

The liver is the second-largest organ in the body after the skin tissue, and it is located between the heart and gastrointestinal tract [37]. It occupies 2–3% of body mass of an adult, and it is responsible of more than 500 kinds of vital functions, including immune response, detoxification, energy metabolism, blood protein synthesis, etc.

The liver is a complicated organ in terms of its microstructures and cellular compositions (Fig. 7.13). It consists of 1.5 million hexagonal subunits called hepatic lobules, and the height and diameter of each lobule are about 2 mm and 1–1.3 mm, respectively [39]. The liver is composed of various kinds of cells, which are hepatocyte (a parenchymal cell for major hepatic functions), cholangiocyte (biliary epithelial cell), hepatic sinusoidal endothelial cell (HSEC), hepatic stellate cell (HSC),

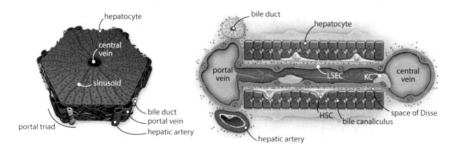

Fig. 7.13 Anatomy of the liver lobule. (Reproduced with permission from [38]). LSEC: liver sinusoidal endothelial cell; KC: Kupffer cell; HSC hepatic stellate cell

Kupffer cell, hepatic stem cell, interstitial cell, etc. As an epithelial cell, hepatocytes show a specific polarity in the liver microstructure, and their apical parts are directed to a vascular structure called a hepatic sinusoid, and the basal part forms bile canaliculi with other hepatocytes to form a lumen by which bile acid flows. Types of cells other than hepatocytes are called non-parenchymal cells, among which HSECs account for a large percentage of the liver after hepatocytes. HSECs are sinusoidal endothelial cells, and they form highly porous endothelium of hepatic sinusoid, enabling exchange of various kinds of proteins, chemicals, and metabolites. In addition to the basic functions of endothelial cells, HSECs also perform the specialized function of involvement in blood coagulation by producing factor VIII [40]. Hepatic stellate cell, also called Ito cell, is a kind of interstitial cell residing perisinusoidal region called space of Disse, and the major functions are storing lipid and regulating liver regeneration [41]. HSC is also a cell that contributes to the progress of cirrhosis by differentiation into myofibroblasts, and it is considered very important for various liver diseases. Kupffer cell is a kind of macrophage residing in the lumen of hepatic sinusoid, and it protects the liver from microbes from gut or outside of body [42]. Cholangiocyte forming intra- or extra-biliary tract, except bile canaliculi, is a type of tubular epithelial cell, and bile acid flows through it to be collected in a gallbladder located outside the liver tissue. In the case of hepatic stem cell, it is essential for tissue/organ maintenance and regeneration, as in other tissue stem cells [43].

Unlike other tissues, ECM of liver accounts for only about 10% of the total liver volume, but it plays a crucial role in the regulation of liver function, repair, and regeneration [44]. Type I, III, and V collagens are the major fibrillar components supporting stiffness of the liver, and type IV collagen is a non-fibrillar one forming a network between ECM components [45]. Besides, type VI and VIII collagens are cytokine-binding collagens regulating tissue repairing or regenerations. In addition, glycoproteins include fibronectin and laminin, and proteoglycan has perlecan and hyaluronic acid. All of these components cover a wide range of regulatory functions in the liver (Table 7.1) [46].

The liver is a central organ for chemical metabolisms, which makes it important in drug developments and the medications. Drugs absorbed in the body are mainly biotransformed in the liver, affecting pharmacokinetic characteristics such as absorption, distribution, metabolism, excretion, and efficacy, and their toxicity can be unexpectedly changed [47]. Therefore, it is very important to predict how drugs are metabolized in the liver, which is verified first through in vitro test and preclinical trials, and then, finally, selected drugs are sold in the market. Despite the process of consuming such large amounts of time and money, more than 800 kinds of the currently marketed drugs are known to be not free from hepatotoxicity [48]. In addition, about 40% of the drugs withdrawn from the market since the 1990s have been withdrawn due to hepatotoxicity, which is about 10% more than the second-largest reason, cardiotoxicity 33% [49]. This is because drug metabolic enzymes related to drug metabolism phases I and II, such as cytochrome P450 family, UDP-glucuronosyltransferases, or sulfotransferases, show a lot of differences between people and other animals, which is fundamentally unpredictable for all drug

Table 7.1 Cellular component of liver tissue

Classification	Cell type	Cell	Function
Parenchymal cell	Epithelial cell	Hepatocyte	General functions of liver (synthesis of blood proteins, detoxification, etc.)
Non-parenchymal cell	Endothelial cell	Hepatic sinusoidal endothelial cell	Formation of sinusoidal microvessel Synthesis of factor VIII
	Hepatic interstitial cell	Hepatic stellate cell	Storage of fat Liver regeneration Progress of fibrotic liver diseases
		Hepatic fibroblast	Maintenance of liver tissue Progress of fibrotic liver diseases
	Macrophage	Kupffer cell	Regulation of immune response Regulation of hepatic diseases
	Epithelial cell	Cholangiocyte (biliary epithelial cell)	Formation of biliary ducts Regulation of bile acid components
	Stem cell	Hepatic stem cell	Liver regeneration Turnover of various liver cells

metabolisms that occur between people and other species [50]. This is not just a problem due to the differences between species, but it is also because it is difficult to find mechanisms or biomarkers for hepatotoxicity [51]. Therefore, it is urgently required to develop an in vitro model to recapitulate human pathological and physiological physiology accurately in both drug development and clinical practice.

7.3.2   3D Bioprinting of In Vitro Liver Models

In developing in vitro liver models, it is very important to implement 3D environment, microfluidic condition, various cell compositions, and extracellular substrate as native liver tissue, and many studies have been reported about each issue. Firstly, it is already well known that 3D environments using various approaches including sandwich culture, spheroid, or scaffolds exhibited matured hepatocyte functions better than 2D in terms of viability, biosynthesis, CYP450 expression, and so on [52–54].

In addition to the 3D environment, attempts have been made to develop a more functional liver model by co-culture of non-parenchymal cells. Wang et al. developed a 3D liver model by co-culturing human-induced pluripotent stem cell (iPSC)-derived hepatocytes and human umbilical vein endothelial cells (HUVEC) [55]. Compared with the experimental group that cultivated hepatocytes alone, the in vitro liver model with HUVEC showed increased albumin secretion and CYP450 expression. Besides, acute and chronic hepatotoxicity caused by acetaminophen (APAP) showed increased drug sensitivity similar to that in native liver. Wei et al.

fabricated a scaffold-based 3D liver model co-cultured with hepatocytes and HSCs, and they found that biosynthetic function and CYP expression were increased [56]. When the CYP450 activity was measured using testosterone and phenacetin, the activity was significantly high in the co-culture model. In the case of Liu, where hepatocytes were co-cultured with HUVEC and fibroblast in the micropatterned fibrous mats, formation of bile canaliculi and capillary tube was significantly enhanced, and, as in previous studies, biosynthetic functions and activity of CYP450 were increased [57]. Thus, co-culture of parenchymal and non-parenchymal cells means more than simply adding different types of cells that perform distinct functions, showing that intercellular interactions improved the functionality of 3D liver constructs.

Considering microfluidic environment for in vitro liver model has important meaning in recapitulation of mechanical stimulus mimicking blood flow, which makes 10–50 mPa of shear stress at hepatic sinusoidal endothelium [57]. For liver models with microfluidic environments, the framework using synthetic polymers such as polydimethylsiloxane (PDMS), polycaprolactone (PCL), or poly (ethylene-vinyl acetate) (PEVA) is essential to create channels or reservoirs in which cell culture media can flow or be stored, and most of them are made in the form of a chip [58]. Various kinds of liver cells can be located in the framework in accordance with native mimicking positions, and they are mechanically stimulated by cell culture medium flowing in uni- or bidirection using pumps or rockers. Toh et al. developed a microfluidic drug testing device called 3D HepaTox Chip, and viability and synthesis of albumin and urea enhanced compared with conventional 3D culture method [59]. Besides, when treated with drugs such as APAP, diclofenac, and rifampin, 3D HepaTox Chip showed in vivo-like drug sensitivity. Rennert et al. recapitulated the sinusoidal structure of the liver, including space of Disse, using hepatocyte and HUVEC microfluidic in devices made of PDMS [60]. By flowing a cell culture medium 50 mPa (0.5 dyn/cm^2) to the endothelium part, native liver-like shear condition was applied, and the biosynthesis of hepatocytes, the formation of bile canaliculi, and the activity of drug metabolic enzymes such as CYP450 and multidrug resistance-associated protein 2 (MRP2) were significantly increased. Moreover, hepatocytes in the dynamic flow condition showed human-liver-relevant morphology forming microvilli where MRP2 is located in the native liver tissue. To implement liver ECM environment, liver dECM is known to be the optimized material. However, the application of liver dECM to liver model is the case of using 3D printing and liver dECM bioink, which is introduced in the following section, and this material has been mainly applied as a graft for regenerative medicine.

The above elements for recapitulating the native mimicking liver can be produced as a platform through 3D printing technology, enabling fabrication of even more similar microstructure of liver tissue compared with conventional techniques. Bhise et al. developed a 3D-printed in vitro liver model by directly printing the hepatocyte spheroids which were encapsulated into the photocurable methacrylated gelatin (GelMA) bioink, on a PDMS-based microfluidic platform using 3D printing and containing hexagonal shape of liver lobule [61]. Lee et al. used collagen and gelatin bioinks for encapsulating and printing cells to produce a one-step fabrication-based liver-on-a-chip containing PCL-based chip body and vascular parts. In particular, in the case of vascular regions, the use of gelatin bioinks led to the formation of "monolayer" that mimics the liver sinusoid structure after printing [62]. In both cases using 3D printing technology, the biosynthetic functions of the liver and viability were improved, and it was easier to fabricate microstructures of native liver tissues. Especially, the approach of Lee et al. has the advantage of being able to easily control the way of compartmentalization of cell according to interest, and more sophisticated cell arrangement was possible.

As shown above, liver ECM plays a significant role in hepatic functions, and Lee et al. developed a decellularized liver-derived bioink to recapitulate liver ECM environments for 3D bioprinting applications (Fig. 7.14a) [63]. After confirming that liver dECM bioink is applicable to 3D bioprinting by its tunability and versatility, it was validated that the biosynthetic functions of the hepatocyte structure printed with liver dECM bioink are significantly improved compared with the hepatocyte structure printed with bioink made using medical grade type I collagen. Skardal et al. reported a different type of liver dECM bioink by mixing gelatin-based photo-curable material, photoinitiator, and various kinds of crosslinkers to improve mechanical property [64]. Compared to Lee et al., where the bioink underwent just thermal crosslinking, the mechanical property of the bioink was more controllable and improved. However, with spheroid of hepatocyte, hepatic stellate cell, and Kupffer cell, functions and viability were not well maintained from 10 days by supposed toxicities from various chemical reactions. Lee et al. applied their liver dECM bioink without any chemical additives to in vitro liver model containing hepatocytes and HUVECs, thereby recapitulating various cell compositions, native-like structures, and microfluidic conditions of liver tissue and developing advanced liver-on-a-chip (Fig. 7.14b) [65]. The liver in vitro model is embodied in flow both of the blood and of the bile acid in the upper and lower parts, respectively, and the hepatocytes and HUVECs are located like native sinusoids. In addition, the maturation of hepatocytes was more induced when bile acid channel was present, and the expressions of CYP450 were also increased. Besides, APAP treatment showed similar drug toxicity to those of native liver.

In conclusion, the in vitro liver model with the precise location of cells and various microfluidic environments of liver tissues could be fabricated through 3D bioprinting technology. Moreover, variety of biochemical environments of liver ECM were implemented at the same time, thereby enabling the development of structurally and functionally native relevant liver tissues.

In terms of toxicity and drug metabolism, liver tissue is of major importance in drug development and clinical trials, and for this reason, development of in vitro liver tissue is currently an indispensable issue. As shown above, many studies have been conducted to implement various elements of what native liver is composed of, and there has been considerable progress through 3D bioprinting and use of liver dECM bioink. Although not discussed in detail in this chapter, considering the limitations of primary hepatocytes that are not able to maintain functional phenotype in vitro, it is thus far very challenging to overcome the immature hepatic functions of cancer cell lines, immortalized cell lines, and stem cell-derived hepatocytes [46].

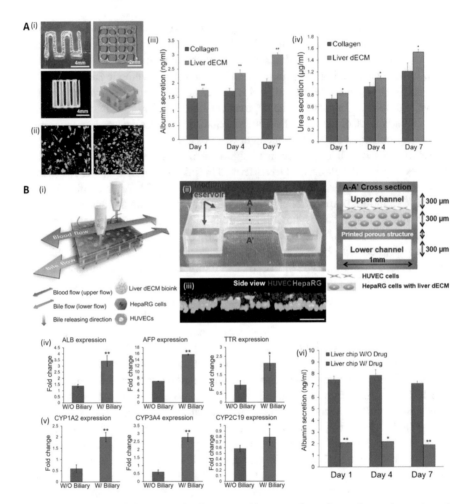

Fig. 7.14 Liver dECM bioink and 3D liver-on-a-chip possessing a liver microenvironment and biliary system. (**a**) (i) Different types of 3D-printed constructs and (ii) high cell viability after bioprinting using liver dECM bioink. Improved biosynthesis of (iii) albumin and (iv) urea in the liver dECM bioink group compared to type I collagen group. (Reproduced with permission from [61]. Copyright 2017 American Chemical Society). (**b**) (i) Schematic of 3D liver-on-a-chip possessing a liver microenvironment and biliary system. (ii) Fabrication of chip body and cross-sectional view of cell part. (iii) Location of each cell component after 3D bioprinting of liver-on-a-chip. Improvement of (iv) functional markers and (v) drug metabolic enzymes in the liver chip with biliary (vi) and in vivo mimicking drug response on it. (Reproduced with permission from Ref. [65])

Moreover, there is a lack of research on how hepatic stellate cells or Kupffer cells, which have a great influence on the regulation of both normal and diseased liver tissue, can be implemented in vitro. If these limitations are overcome in the future and eventually applicable to the multi-organ culture system to be considered, the in vitro liver model will contribute greatly to reducing various social costs regarding healthcare.

7.4   Kidney

7.4.1   Structural and Physiological Functions of the Kidney

Kidneys are a pair of bean-shaped organs located on the right and left side in the retroperitoneal space of all vertebrates and well protected by muscle, fat, and ribs [66]. They are a highly vascularized, multifunctional organ that is responsible for an excretory role in the removal of wastes, metabolites, and toxins from the blood through the formation of urine. In addition, other vital functions of the kidney include (a) maintaining homeostasis by regulating the extracellular fluid levels, blood pressure, electrolyte, and inorganic ion balance through producing hormones and other factors to retain the internal body environment consistent, (b) gluconeogenesis, and (c) synthesis of erythropoietin, renin, 1,25-dihydroxycholecalciferol hormones, etc. [66]. The length, width, and thickness of the kidney are about 11–14 cm, 6 cm, and 4 cm, respectively. Male's kidney weight (127–170 g) is slightly higher than the females'. Blood flows through paired renal arteries in the kidney and exits through the paired veins. The urine is transported through the muscular tube named ureter into the bladder. Kidneys are directly surrounded by tight, fibrous capsule including dense connective tissues that serve as protection to the kidney and grip their shape (Fig. 7.15a). Frontal view of the kidney reveals that the kidney structure is divided into the outer renal region (renal cortex) and inner renal region (central medulla). Renal columns are the extension of the cortex connective tissue down toward the renal medulla to separate most specific structures of the medulla (i.e., renal pyramids and renal papillae). The renal columns also divide the kidney into six to eight lobes and offer a supporting frame to vessels that flow in the renal cortex. Renal hilum is the region of the kidney through which ureter, renal artery and veins, nerves, and lymphatics enter and leave.

The function of the kidney is executed by the action of many different types of cells present in a million tiny filters called "nephrons." Each microscopic filtering unit (nephron) is composed of renal corpuscles and concomitant renal tubule of adjoining segments. The tubule, which drains the urine into the ureter, contains several segments such as proximal convoluted tubule, the loop of Henle, the distal tubule, and collecting duct (Fig. 7.15b). The renal corpuscle is the first filtering part of the functional nephron which comprises Bowman's capsule, blind-ended tubule composed of a monolayer of epithelial cells, and the glomerulus, tuft of glomerulus capillaries which contain the strict arrangements of the fenestrated monolayer of glomerular endothelial cells, podocyte, around the glomerulus basement membranes (Fig. 7.15c, e). The inner part of the glomerulus capillary is a fenestrated monolayer of endothelial cells; and the middle layer is the glomerulus basement membrane consisting of the meshwork of fibers such as type IV collagen, proteoglycans, and glycoproteins. The glomerulus basement membrane is the trilaminar glomerular basement membrane consisting of the thick middle layer lamina densa, the lamina rara interna, and the lamina externa which support the outermost podocyte layer of glomerulus capillary (Fig. 7.15d, e). Ultrafiltration of around 20%

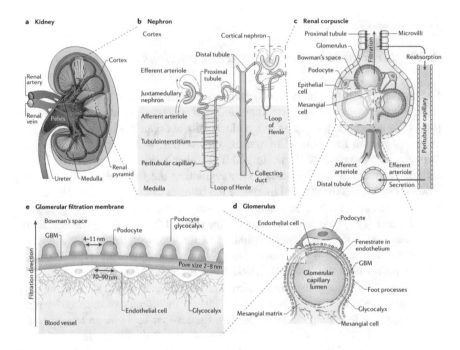

Fig. 7.15 Kidney anatomy and function: (**a**) Kidney consisted of three parts such as renal cortex, renal medulla, and renal pelvis. Each kidney comprises of many pyramids. (**b**) The functional unit of the kidney is called "nephron," and it is divided into renal corpuscle (glomerulus and Bowman's capsule) and the tubular parts such as proximal tubule, the loop of Henle, distal tubule, collecting ducts, and peritubular capillaries near the renal tubules. (**c**) The renal corpuscle cross-sectional view demonstrates the afferent arterioles, which carry the blood to the glomerulus region. The glomerulus capillary regulates the filtration of solute from the blood into the Bowman's capsules and further filtered blood transported by the efferent arterioles. Filtered solutes (wastes) in Bowman's capsules space travel to proximal tubules region, where reabsorption of nutrients takes place. The proximal tubule is the monolayer of epithelial cells, which contains densely packed microvilli. Furthermore, the next part is the distal tubule, which is composed of a monolayer of the epithelial cells that do not comprise the microvilli. Filtration of blood takes places in the glomerulus; solutes reabsorption follows in proximal tubules and transports to nearby peritubular capillaries; and secretion happens in between the distal tubule and adjacent peritubular capillaries. (**d**, **e**) The glomerular filtration barrier composed of the monolayer of fenestrated glomerular endothelial cells, glycocalyx, glomerulus basement membrane and podocytes cells, and mesangial cells is present in between the glomerulus capillary to provide the structural support to the glomerulus capillary networks. (Reproduced with permission from [66])

blood from the glomerulus to Bowman's capsules is directed through the endothelium, basement membrane, and podocytes layer of glomerulus capillary by acting as selective filtration barrier. The podocytes are a specialized epithelial cell type (footprint-like structure) of the glomerulus filtration barrier that wraps around glomerular capillaries. The filtration slit between the foot processes restricts the plasma proteins from entering the glomerulus ultrafiltration and further plays a role in podocyte signaling. Glycocalyx is a negatively charged glycoprotein that wraps the foot processes and slit diaphragm that provides negative surface charge all over the glomerulus filtration barrier, sustaining the podocytes architecture through electrostatically repelling nearby foot processes. Mesangial cells are smooth muscle-like cells present in between the glomerulus capillary to provide the structural support to the glomerulus capillary networks. Contractile activity of the mesangial cells is also assumed to impact on the glomerulus filtration function by decreasing the capillary surface and permeability. After ultrafiltration of the blood, the glomerulus filtrate accumulates in Bowman's capsule and is further transported to the next renal tubule segment of the nephron. The renal tubule region is composed of the cuboidal monolayer of the epithelial cells that provide the reabsorption and removal of solutes or molecules from glomerulus filtrate.

The proximal convoluted tubule is responsible for body homeostasis through the reabsorption of salts, water (60–70%), and nearly all nutrients. Proximal convoluted tubule segments encompass apical brush border, which contains higher numbers of microvilli, mitochondria, Golgi apertures, and more basolateral membrane invaginations and hence holds a large capacity of reabsorption. After the renal proximal convoluted tubule segment, the next continuation segment of the renal tubule is the loop of Henle, covering a thin descending limb, thin ascending limb, and thick ascending limb that provides reabsorption and the establishment of high osmolarity of the renal medullary interstitial to concentrate and dilute urine. The terminal portion of the nephron is the distal convoluted tubule and collecting duct, which is involved in the reabsorption of water as needed to yield urine at a particular concentration that preserves fluid homeostasis of the body. The stroma tissues and vasculature interact with the cells of the nephron to provide the overall complexity and kidney function. In addition, the removal of the uremic toxins governs by the transporter proteins exist in renal tubule cells.

7.4.2 3D Bioprinting of Advanced Renal Tubular Tissues

The outstanding efforts have been explored for the development of in vitro kidney tissue models using various approaches such as a hollow fiber, 2D models, and 3D gels and microfluidic kidney-on-chip models. Most of the fabricated devices can culture only one type of cells in the microchannels, and therefore these models cannot reflect the cellular heterogeneity of the kidney. Although the development of a fully functional kidney model is still in its infant stage, in recent years, researchers focus on 3D bioprinting technology as an alternative fabrication method that potentially offers the benefits for the creation of advanced in vitro tissue models.

For the first time, bioprinting of 3D-convoluted renal proximal tubules on perfusable chips was developed by Homan et al. using sacrificial-based 3D bioprinting methods (Fig. 7.16) [21]. They printed the sacrificial pluronic-F127 ink on the surface of coated gelatin–fibrinogen ECM in the chamber of the chip and further poured the additional gelatin–fibrinogen ECM bioink around the printed fugitive pluronic-F127.

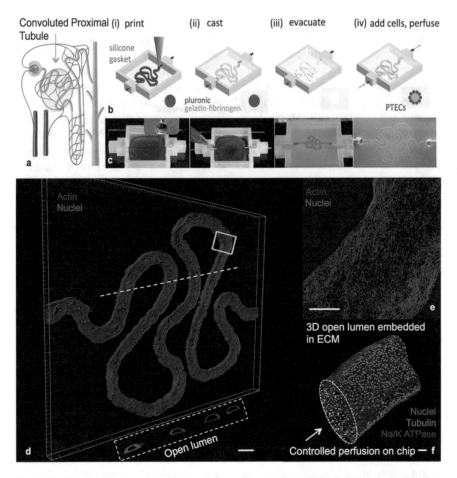

Fig. 7.16 3D-convoluted renal proximal tubule on chip. (**a**) Illustration of a nephron emphasizing the renal proximal convoluted tubule. (**b, c**) Schematic illustration of the fabrication steps in the construction of 3D perfusable renal proximal convoluted tubules, in which a sacrificial ink is first printed on a gelatin–fibrinogen extracellular matrix (ECM) (i), after which further additional ECM is poured around the printed structure (ii), the sacrificial ink is evacuated to generate an open tubule (iii), and renal proximal convoluted epithelial cells (PTECs) are seeded inside the tubule and perfused for long time periods (iv); (**d**) a 3D image of the printed convoluted proximal tubule is obtained by confocal microscopy (actin and nuclei are stained with red and blue, respectively); cross-sectional view revealed below where PTEC cells demarcate the open lumens in 3D, scale bar = 500 μm; (**e**) higher-magnification image of the region in (**d**) represented by the white rectangle, scale bar = 200 μm; (**f**) 3D rendering renal proximal convoluted tubule, where an open lumen confined with an epithelial lining is directionally perfused on chip; Na/K ATPase, acetylated tubulin, and nuclei are stained in red, orange, and blue, respectively, scale bar = 50 μm. (Reproduced with permission from [21])

A convoluted perfusable channel was eventually generated in ECM hydrogel by removing the sacrificial pluronic-F127; subsequently, the human proximal tubule epithelial cells (hPTEC-TERT1 cell line) were seeded inside the channel. Perfused proximal tubule cells in this model formed a tight monolayer of epithelium after

2 months of culture. The epithelial barrier function was evaluated by perfusion of fluorescein isothiocyanate (FITC)-labeled inulin. Moreover, the proximal tubule cells in this system showed higher albumin uptake and cellular morphology compared to two-dimensional (2D) culture. Overall, the perfusable proximal tubule-on-a-chip maintained its functionality over 2 months, signifying that 3D bioprinting could permit the fabrication of complex kidney structure with superior functionality and durability. However, this model has not been able to mimic the in vivo microenvironment of the vascularized proximal tubule due to lack of adjacent peritubular capillary; in the in vivo condition, peritubular capillary exists around the tubules to provide the tubule–vascular interaction.

To overcome this challenge, a further study of this group has bioprinted the perfusable 3D-vascularized proximal tubule on a chip (3D VasPT) using modified gelatin-fibrin bioinks and fugitive ink (a combination of 25 wt% pluronic and 1 wt% PEO-poly(propylene)-PEO) (Fig. 7.17) [67]. The bioprinted 3D VasPT exhibited

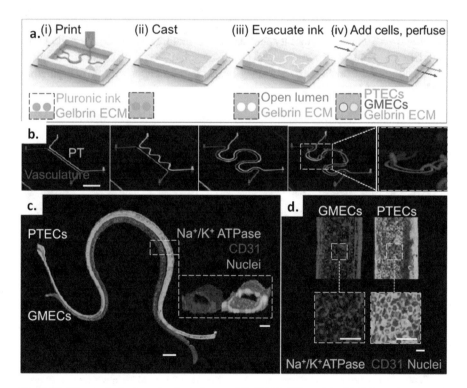

Fig. 7.17 3D vascularized proximal tubule on a chip (3D VasPT) models. (**a**) Schematic representation of the fabrication steps of 3D VasPT. (**b**) Fabrication of various designs (simple and complex) of 3D VasPT. (**c**) Immunofluorescent staining of the 3D VasPT model, where Na+/K+ ATPase, CD31, and nuclei are stained in green, red, and blue, respectively (scale bar: 1 mm); cross-sectional view of the two lumens (PTEC and GMECs) (scale bars: 100 μm). (**d**) High-magnification images of 3D VasPT tissue after staining (scale bars: 100 μm). (Reproduced with permission from [67])

selective renal reabsorption through tubular–vascular channels interaction. Longtime culture of the proximal tubule (PT) with the presence of adjacent vasculature on 3D VasPT in perfusion condition was assessed. Their findings achieved a confluent monolayer of PT epithelium, which significantly improved in their microvilli length and density. Additionally, the hyperglycemic states were also induced on the 3D VasPT to demonstrate that this platform can be utilized to explore relevant interactions for diseases (e.g., diabetes).

Remarkably, bioinks such as fibrinogen, gelatin, thrombin, and PEG used in the above studies are not the main kidney-specific ECM components. Although these bioinks compositions were partly satisfying the mechanical characteristic of the kidney tissue, they lack kidney-specific biochemical composition that regulates viability, proliferation, and differentiation of the cells. In this regard, developing alternative bioinks that can satisfy both the kidney-specific biochemical compositions and mechanical properties is required. Recently, Ali et al. developed the photocrosslinkable kidney decellularized ECM bioink (KdECMMA) for 3D bioprinting in vitro kidney tissues (Fig. 7.18) [68]. Renal tissue constructs were printed using kidney cell-laden KdECMMA bioinks. The bioprinted cell-laden-KdECMMA construct showed increased viability and enhanced maturation of kidney cells compared to the control group (GelMA). This result reflects that the biochemical composition of the decellularized kidney ECM plays an important role in regulating the behaviors of kidney cells.

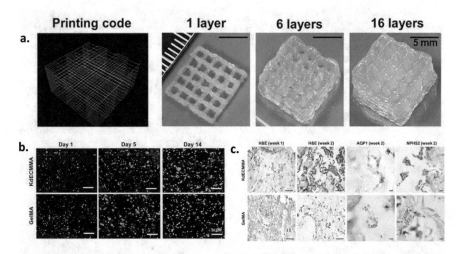

Fig. 7.18 3D bioprinting of renal tissue construct using kdECM-based bioinks. (**a**). Printing code and optical images of the printed KdECMMA-based constructs. (**b**) Live/dead images of bioprinted renal constructs and control group (GelMA) (green, live; red, dead). Scale bar = 50 μm. (**c**) H&E images showed the human primary cells organization into glomerular- and tubular-like structure in the printed renal constructs after 2 weeks' culture, scale bar = 50 μm. Immunohistochemical staining for AQP1 and NPHS23 revealed the preservation of kidney-specific phenotype in the bioprinted renal constructs compared to the control group (GelMA), scale bar = 50 μm. (Reproduced with permission from [68])

7.5 Heart

7.5.1 Structural and Physiological Functions of Hearts

The heart, which has the function of pumping the blood into the arteries that transport oxygen and nutrients throughout the organs of the body, is the central part of the circulatory system [69, 70]. It is situated at the middle of the chest, underneath slightly left of the breastbone (sternum). It is the most hardworking muscular pump in the human body, as it beats continuously. Due to this crucial role, the heart can be one of the most vital organs in the human body; even minor abnormalities in this organ may cause severe effects in the body. Working functions of the muscular heart are made possible by the operating principle of the four chambers that collect and transport the oxygen-poor or oxygen-rich blood (Fig. 7.19) [69].

The two chambers at the top portion of the heart are recognized as right and left atria, and the other two chambers positioned in the bottom portion of the heart are known as ventricles. Atria receive the oxygen-deficient blood that is returning from the various parts of the body, whereas ventricles pump oxygen-rich blood to the organs of the body. The atria chambers are separated by the atrioventricular valves, which are comprised of the tricuspid valve on the right and mitral valve on the left. Similar to atria, the two ventricles chambers are also divided by the semilunar valves (pulmonary and aortic). The wall of the heart is composed of three layers: the

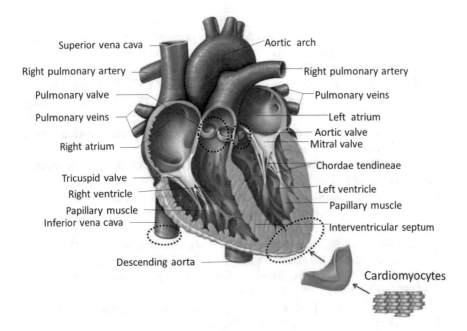

Fig. 7.19 Illustration of cardiac anatomy. (Reproduced with permission from [69])

outer tough fibrous thin layer that is made up of connective tissue is called "epicardium," the thick middle layer consisting of cardiac muscle fibers is "myocardium," and the thin innermost layer is "endocardium," containing squamous epithelium, which covers the heart chambers and valves. The thick-layer myocardium of the heart wall has intricate patterning of cardiac muscle, which is enclosed by the collagen framework. The intricate patterning of the cardiac muscle is due to swirling and spiraling of muscle cells around the heart chambers. This intricate patterning permits the heart to more effectively pump the blood.

The cardiac muscles contain two kinds of cells: muscle cells, which are readily capable of contracting, and pacemaker cells of the intrinsic conduction system. These cardiac muscle cells interlinked through intercalated discs, which provide the quick transmission to electrical impulse of the action potential from the pacemaker cells. The complex structure of the heart facilitates blood transport into the different organs of the body and its return to the heart. The vessel that transports the blood into the heart is known as veins; those vessels which transport the blood from the heart to the different parts of the body are called arteries. The largest artery arises from the left ventricles, called the aorta, and it further branches into two iliac arteries. With consistent functioning, the heart is capable of sufficiently transporting the oxygen and nutrients into different parts of the body.

7.5.2   3D Bioprinting of In Vitro Cardiac Tissue Models Using Heart dECM Bioink

Worldwide, cardiovascular disease (CVD) is the primary cause of heart failure and mortality. In 2008, approximately 86.2 million adults in the USA were diagnosed with at least one type of CVD, and it is estimated that around 40.5% of the total American population by 2030 will have some kind of CVD [71]. According to updated data of the American Heart Association 2017, 7.9 million adults suffer myocardial infarction (MI) out of a total 16.5 million adult population with coronary heart disease [72]. The progress of coronary heart disease can cause heart failure and death, perhaps due to blockage of the coronary artery, which leads to the progressive deterioration cardiomyocyte by necrotic and apoptotic environments [72]. Therefore, at end-stage heart failure or congenital heart anomalies, the patient frequently requires alternate options of treatment to improve the prognosis. The currently available treatments for end-stage cardiac failure are the cardiac transplantation or left ventricular assist devices (LVADs). However, the cardiac transplantation is limited due to several shortcomings such severe donor shortages [73], adverse immune rejection after transplantation [74], and need of lifelong immunosuppression therapy; in addition, LVADs devices support the cardiac function for a limited time [75]. Furthermore, 31% of drugs have been withdrawn in the USA due to drug-induced cardiac toxicity [76]. This necessitates in vitro cardiac models for transplantation and also for drug-induced cardiac toxicity screening.

Cardiovascular tissue engineering has been recognized as an alternative approach developing the engineered cardiac tissues for the transplantation and drug screening [77]. For the cardiac tissue engineering, choices of the appropriate key elements such as proper cell source, advanced biomaterials mimicking the cardiovascular ECM, vasculature, physiologically important contractile function, and electrome-chanically coupling cues are crucial [78]. The selection of the biomaterials for the fabrication of in vitro cardiac tissue is a vital factor, as it must recapitulate physical and biochemical properties of the inherent tissue ECM [79, 80]. In this regard, the use of decellularized ECM derived from the native myocardium may inherit the natural ECM compositions. As a result, the cell-biomaterials interaction can be promoted for better engineering relevant cardiac tissue model. However, recapitulation of the complex heterogeneous structure of the native cardiac tissue still remains challenging [81].

Therefore, innovative methodologies of bioengineering heart models are explored to tackle these issues and further expand prospective of cardiac tissue engineering. Recently, 3D bioprinting has been introduced as a promising technology in cardiac tissue engineering by enabling the deposition of biomaterials and cells in predefined area [82–84]. In this chapter, we discuss the necessity of in vitro cardiac tissue models and highlight the 3D bioprinting of cardiac tissue models using dECM-based bioink.

Since decades, several standard conventional methods such as decellularization and recellularization [85], heart decellularized extracellular matrix (hdECM) scaffolds [81], hdECM sheet [86] electrospun fiber mat [87], and hydrogel of hdECMs [11, 88] have been employed in cardiovascular tissue engineering and showed the encouraging outcomes. However, they have dimensional discrepancy [89]. In addition, these types of models need the post cell seeding process; the deficiency of appropriate microgeometry usually results in imperfect distribution and infiltration of the cells in the inner areas. Additionally, without the vasculature network, oxygen diffusion, and nutrition to the infiltrated, cells can be limited inside the implant and which may result in the cell death [90]. Therefore, to better recapitulate structural arrangement of native tissue, currently, 3D bioprinting has been considered as a promising technique. To mimic the function and structural complexity of the native tissue, bioprinting an in vitro tissue model through layer-by-layer deposition of the cell-laden tissue-derived dECM in an organized manner is crucial [82, 83]. 3D bioprinting of in vitro model enables the construction of 3D porous structures, allowing efficient transport of oxygen and nutrition [91].

Specifically, the development of in vitro cardiac model demands the innovative processes such as the use of 3D bioprinting with hdECM, proper cell source (human-induced pluripotent stem-cell-derived cardiomyocytes [hiPSC-CMs]), and complex vascular network. The balance arrangement of these parameters is able to provide the advanced in vitro model of vascularized cardiac tissue with enhanced functional behavior. Fabrication of in vitro cardiac tissue model using 3D bioprinting and dECM-derived bioinks represents the attracting areas where the dECM bioink signifies the innovative class of material. Overall, during bioprinting, the bioink has to show the shear-thinning behavior so that it can smoothly dispense with low pressure, and after extrusion, the bioink should rapidly crosslink to maintain the structural integrity of the printed construct with high cell viability and functionality [92]. In general, the dECM bioink showed the temperature-sensitive sol–gel transition behavior, exhibiting sol state at low temperature and crosslinked to gel state at body temperature (37 °C) [93]. By considering all together criteria as mentioned above, Pati et al. developed the idea of 3D bioprinting of dECM bioinks derived from harvested heart, cartilage, and fat tissues. Cell-laden dECM bioinks were printed into the porous PCL-based construct (Fig. 7.20) [93].

Furthermore, Das et al. [94] engineered heart tissue (EHT) using cardiomyocyte-laden hdECM and 3D bioprinting technology to demonstrate the effect of microenvironments of bioinks and culturing conditions on cardiomyocyte maturation (Fig. 7.21). The printed constructs were cultured in the static and dynamic culture condition. The results showed the dynamic conditioning significantly promoted the structural arrangement of cardiomyocytes and the differential gene expression patterns. In addition, qualitative and quantitative analysis reflected the improved cardiomyocytes maturation in hdECM compared to collagen under the same culture conditions.

Furthermore, Jang et al. attempted multicellular patterning of the prevascularized cardiac patch using cell-laden (human cardiac progenitor cells and MSCs)-porcine hdECM bioink (Fig. 7.22) [95]. 3D patterned prevascularized cardiac patch encouraged host–graft anastomosis and tissue formation after in vivo implantation in the rat model of myocardial infarction (MI). The developed patch showed robust vascularization, enhanced cardiac function, reduced cardiac hypertrophy, and ameliorated fibrosis.

However, printed constructs with hdECM bioink were unable to convincingly maintain the structural fidelity during the printing process due to slow gelation of the bioink. To improve the printability and fidelity of the hdECM

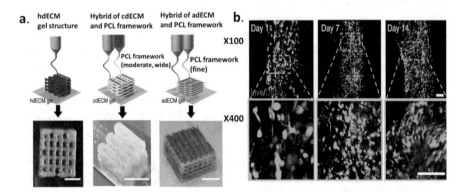

Fig. 7.20 3D bioprinting process of dECM-based bioinks. (**a**) Printing process of specific tissue structure with dECM-based bioinks such as heart, cartilage, and adipose tissue constructs was bioprinted with heart dECM (hdECM), cartilage dECM (cdECM), and adipose dECM (adECM), respectively, and in combination with PCL supportive framework (scale bar, 5 mm). (**b**) Representative live/dead images of the constructs (scale bar, 100 μm). Cell viability was >95% at day 1 and >90% at both day 7 and day 14. (Reproduced with permission from [93])

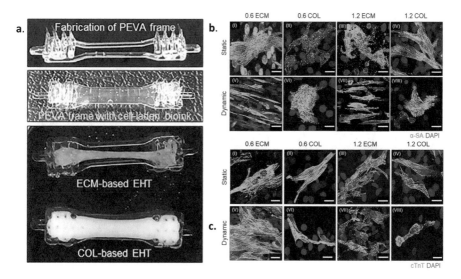

Fig. 7.21 Bioprinting of engineered heart tissue (EHT). (**a**) 3D-printed engineered heart tissue using ECM (hdECM)- and COL-based bioinks. (**b**) Confocal microscopy images of immunostaining for α-sarcomeric actinin (α-SA) and (**c**) cardiac troponin T (cTnT) synthesized by cardiomyocytes in 0.6% and 1.2% ECM and COL cultured statically and dynamically at day 14. Scale bar: 20 μm. (Reproduced with permission from [94])

bioink, several methods were suggested by enhancing its viscosity, reducing its gelation time, or both [96].

As an example, Jang et al. developed the photo- and thermal-crosslinkable hdECM bioinks using vitamin B2 (VB2) and UVA irradiation (Fig. 7.23a, b) [96]. The mechanical properties of hdECM bioinks were enhanced by the sequential thermal and photochemical crosslinking (Fig. 7.23c, d). In particular, mixing of 0.02% VB2 with the 2% hdECM bioink showed remarkable increase in viscosity when compared with hdECM bioink only. However, after thermal and photochemical crosslinking, the stiffness of the VB2-blended hdECM bioink enhanced approximately 33 times compared to pure hdECM. This enhanced stiffness is close to that of native cardiac tissue. Using this photo- and thermal-crosslinkable hdECM bioink, multilayered cardiac construct was successfully bioprinted with enhanced structural integrity (Fig. 7.23e). In addition, higher stiffness of the VB2 mixed hdECM bioink rather reduced the viability and limited cardiac differentiation of progenitor cells (Fig. 7.23f–h).

Similarly, Yu et al. developed photocrosslinkable heart dECM bioinks to fabricate the tissue-specific patterned constructs with highly controlled geometries and instantaneous fine-tuning of the mechanical properties [97]. Their bioprinting process allowed the fabrication of complex tissue constructs (hiPSC-derived cardiomyocyte-laden heart dECM and hiPSC-derived hepatocyte-laden liver dECM constructs) to promote the cellular organization and offer the inherent tissue-specific biochemical microenvironment for high viability and maturation of hiPSC-cardiomyocytes and hiPSC-hepatocytes.

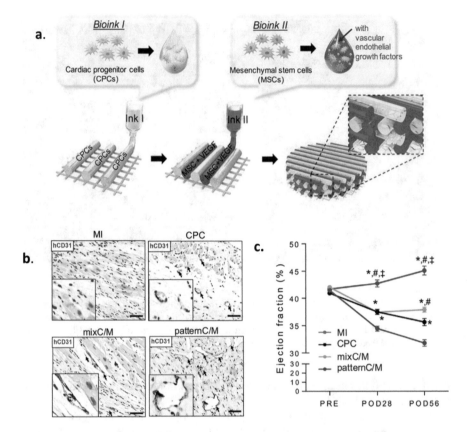

Fig. 7.22 3D bioprinting of multicellular patterned prevascularized cardiac patch using a porcine hdECM bioink. (**a**) Schematic representation of 3D bioprinting process of prevascularized stem cell patch. (**b**) Immunohistochemistry results against human-specific CD31 specificity at infarct regions. (**c**) Effects of prevascularized stem cell patch on the therapeutic efficacy post-MI; EF values at baseline and after 4 and 8 weeks. Error bars represent standard errors of the mean (SEM) ($*p < 0.05$ compared with MI; $\#p < 0.05$ compared with CPC; $\ddagger p < 0.05$ compared with mix C/M). C/M: both CPC and mesenchymal stem cells (MSCs); CPC: cardiac progenitor cells; EF: ejection fraction; MI: myocardial infarction; POD: postoperation day. (Reproduced with permission from [95])

7.6   Airway

7.6.1   Structural and Physiological Functions of Airway

Airway is a part of the respiratory system forming continuous and branched channels that connect from the nose and mouth to the lungs for airflow. It is also called simply the conducting airway, which is discriminated by its function of simply flowing the air instead of exchanging gas in the respiratory system [98]. The airway is divided into two parts, upper and lower airway, according to anatomical position.

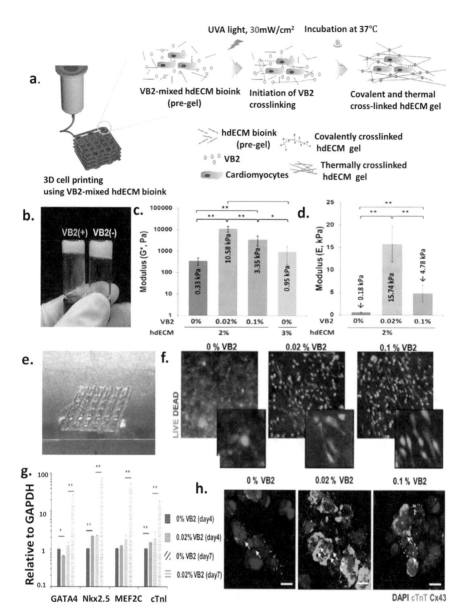

Fig. 7.23 Development of photocrosslinkable, mechanically robust heart decellularized extracellular matrix bioink (hdECM) for 3D bioprinting. (**a**) Representative schematics depicted the two-step crosslinking (vitamin B2-induced UVA crosslinking and thermal crosslinking) of heart decellularized extracellular matrix (hdECM) bioink. (**b**) Gelation behavior of hdECM bioink with (+) or without (−) vitamin B2 (VB2). (**c**) Dynamic complex modulus of each bioink at 1 rad/s depending on the VB2 and bioink concentration (∗: $p < 0.05$, ∗∗: $p < 0.005$). (**d**) Compressive modulus of each bioink at 20% strain depending on the VB2 concentration (∗∗: $p < 0.005$). (**e**) 3D-bioprinted 10 layers of bioconstruct. (**f**) Live/dead images of printed cardiac progenitor cell-laden construct after 24 h; cell viability after 24 h was >95% (200× magnification). (**g**) qRT-PCR analysis of cardiac differentiation-related mRNA expression at days 4 and 7 (error bars represent s.d., ∗: $p < 0.05$, ∗∗: $p < 0.005$). (**h**) Immunofluorescence staining against cardiac troponin T (cTnT, green) and connexin 43 (Cx43, red) at day 7 (scale bar = 10 μm; white arrows indicate the expression of Cx43 at the cell–cell junction). (Reproduced with permission from [96])

The upper airway consists of nasal cavity, pharynx, and larynx in turn from the top, followed by lower airway, which consists of trachea, bronchi, and bronchiole. During respiration, when the air enters the nose or mouth, it goes down through the pharynx to the larynx, which is covered with a small membrane called the epiglottis, preventing food from entering the airway and obstructing breathing. In this chapter, we will cover more about the lower airway, which is closely relevant to respiratory diseases such as asthma or chronic obstructive pulmonary disease (COPD).

Most of the lower airway consists of several concentric layers, including mucosa and submucosa (Fig. 7.24) [100]. The primary function of the mucosa is to protect the respiratory system through mucociliary clearances against foreign substances and chemicals. The mucus secreted from mucosa is composed of water, various ionic substances, and glycoproteins, which lubricates, humidifies, and defends the surrounding tissues.

The airway is a complicated tissue composed of various kinds of cells for each layer. Epithelium, which is located in the innermost and exposed to the lumen, consists of ciliated epithelial cells, mucous cells (goblet cells), serous cells, basal cells, and Clara cells [101]. Among them, the epithelial cells mainly function as a physical barrier to the outside by forming the tight junctions and block the unintended exposures of molecules into the respiratory tissues. Besides, the epithelium performs unidirectional ciliary clearance toward the upper part using cilia, and pathogens or the dusts are discharged via the process with mucus. Both goblet cells and serous cells are epithelial cells, and the former is composed of electron-opaque granules and mainly secretes mucin [102]. On the other hand, the serous cells are composed of electron-dense granules and are reported to secrete water or antimicrobials. Clara cells, also known as club cells, or bronchiolar exocrine cells, also function to protect airway by secreting various substances and are known to secrete glycosaminoglycan

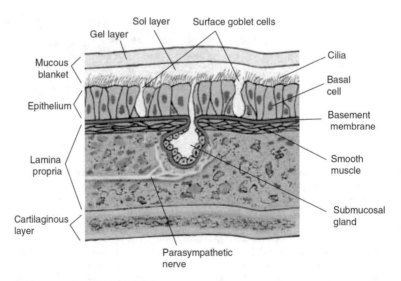

Fig. 7.24 Anatomy of the wall of conducting airway. (Reproduced with permission from [99])

or lysozyme [103]. The basal cell is a kind of progenitor cell for maintaining diverse cell populations in airway epithelium. The basement membrane is an acellular ECM layer lining between epithelium and lamina propria. It not only plays a role of physical barrier but also participates in the maintenance and function of cellular components in the epithelium, thereby regulating the condition of the airway [104]. The basement membrane consists of a type IV collagen and type V laminin secreted from epithelial cells and types III and V collagen and fibronectin produced by the fibroblast of lamina propria [104, 105].

Lamina propria is a loose connective tissue with microvessels and lymphatic vessels lining below the epithelium. The cell component of lamina propria is characterized by fibroblast and various immune cells, which support and regulate conditions of the epithelium [106]. Notably, it has a significant role in mucosal immunity by mucosa-associated lymphocytes producing mainly IgA antibodies to protect airways [107].

Submucosa is a dense connective tissue, and smooth muscle layer is lining in the submucosa. Smooth muscle part is not observed in the trachea, but gradually increases downward and disappears again near the alveoli, and it plays a role in regulating the diameter of the airway against the air pressure generated during breathing. In addition, there are blood vessels and lymph nodes, which support the function of mucosa. There is also a submucosal gland composed of gland serous cells that secrete mucus to the lumen surface, along with the goblet cells of the epithelium [108]. In the case of cartilage ring, it is observed dominantly as the diameter goes up to the wide upper side and gradually decreases as the diameter becomes narrow, disappearing from the terminal bronchioles.

7.6.2 3D Bioprinting of In Vitro Airway Model

As discussed above, airway is a tissue that is directly exposed to the harmful or toxic environments and is susceptible to various pathological conditions. At least three billion people are exposed to the risk of airway damage caused by direct or indirect smoking, and approximately one billion people are suffering from various indoor and outdoor air pollutions and allergens [109]. As a result, with respect to airway diseases, the number of asthmatic patients is 3.34 million, especially in COPD, which is the third highest mortality rate in the world with three million deaths annually. In terms of developing drugs to treat such airway-related diseases, it costs about $ 1.1 billion per a drug, which is more than the cost for cancer or neurological diseases [110]. On the other hand, the probability of entering the market is very low, at 3% compared to other diseases of 6–14% [111]. The difficulty is mainly from a lack of understanding of physiology and pathology about the airway and limitations of animal models used in drug development. The administration route of drugs to airways is also a major contribution of the problem, which is an inhaled route possessing accessibility to the disease site and prevention of systematic side effects [112]. However, considering the difficulty to develop suitable formulation itself for inhaled therapy and toxicity to respiratory tissues, it is a significant obstacle to the drug development. Indeed, approximately 30% of trials for developing inhaled therapy fail due to unexpected toxicity to the respiratory tract observed in later phases in the process. To this point, a human airway model is desperately required to understand pathophysiology of the airway precisely and to reduce compound attrition at late phase by precisely predicting the toxicity and efficacy of drug candidates in the early phase.

Emulation of in vivo relevant airway model requires 3D culture condition, various cellular components and their interactions, lamina propria to support epithelium and to implement microfluidic condition by functional vascular structure thereof, and biochemical environment of ECM [113]. In this section, we will look at the functionality of these requirements in developing more sophisticated in vitro airway models. Firstly, it is well reported that 3D culture is better for implementing the native-like maturity of cells than 2D culture. Fessart et al. confirmed the importance of 3D environment by culturing human bronchial epithelial cells as a 3D spheroid, achieving upregulation of respiratory tract gland markers such as mucin 5B (MUC5B), zinc-alpha-2-glycoprotein 1 (AZGP1), and β-tubulin IV [114]. In addition, 3D spheroid culture improved epithelial polarization, cell–cell interactions, and formation of in vivo relevant lumen architecture. Given that the epithelium of airway is exposed to the air, Whitcutt et al. reported that using a biphasic chamber system is essential for recapitulating such air–liquid interface to maturate ciliary and polarity maturation of epithelial cells in vitro [115].

Myerburg et al. cultured human bronchiole epithelial cells and fibroblasts together using transwell culture system, and, in the presence of subepithelial fibroblasts, the polarity and ciliary phenotype of epithelial cells were more prominent [116]. To create a more complicated 3D airway model, Blume et al. applied both transwell co-culture and microfluidic system and positioned airway epithelial cells and endothelial cells on the permeable membrane (Fig. 7.25a (i)) [117]. The co-cultured model showed higher transepithelial resistance (TER) than other single-cell cultured groups, confirming in vivo relevant barrier functions (Fig. 7.25a (ii)). Besides, the distribution of cells in the co-culture model was more even, as a monolayer, which was more similar to the airway tissue (Fig. 7.25a (iii)). Benam et al. developed a small airway-on-a-chip containing multiple cell types, air–liquid interface, and microfluidic conditions, and chip body was fabricated by stereolithography using polydimethylsiloxane (PDMS) (Fig. 7.25b (i)) [118]. In the fabricated chip, it was confirmed that the pseudostratified airway epithelium and endothelium were formed on and beneath of the porous membrane, respectively (Fig. 7.25b (ii)). In addition, the formation of epithelium and endothelium was validated by observing zonula occludens-1 (ZO1) (Fig. 7.25b (iii)) and platelet endothelial cell adhesion molecule (PECAM) (Fig. 7.25b (iv)), respectively. Moreover, maturation of cilia was determined by β-tubulin IV (Fig. 7.25b (v, vi)), and its function was validated by forward- and return-stroking of a cilium (Fig. 7.25b (vii)).

The airway smooth muscle layer is also critical for airway disease progression, and it is the main effector of narrowing airway in pathological conditions [119]. Recapitulation of its tone is a key element, but difficulties such as in vitro culture of

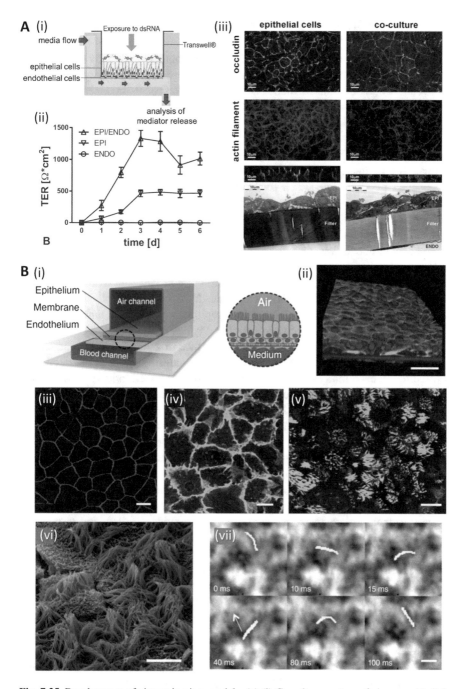

Fig. 7.25 Development of airway in vitro models. (**a**) (i) Co-culture system of airway epithelial cell (AEC) and endothelial cell (EC) using transwell and microfluidic system and (ii) more in vivo relevant transepithelial resistance (TER) in co-cultured group. (iii) Homogeneous monolayer-like cell distribution in the co-cultured model. (Reproduced with permission from [117]). (**b**) (i) Schematic of small airway-on-a-chip and (ii) formation of pseudostratified airway epithelium and endothelium. Immunostaining of (iii) zonula occludens-1 (ZO1) for epithelium and (iv) platelet endothelial cell adhesion molecule 1 (PECAM-1) for endothelium. Observation of matured cilia in epithelium by (v) immunostaining of β-tubulin IV and (vi) scanning electron micrography. (vii) Sequential frames of cilia beating. (Reproduced with permission from [118])

the smooth muscle cell and lack of mechanical environment as in body have become a major obstacle to model development [120]. There have been very few reports about airway smooth muscle model. Among them, West et al. produced a 3D multi-cell microtissue culture model using PDMS well and airway smooth muscle cell/fibroblast-laden collagen hydrogel [121]. In response to potassium chloride and cytochalasin D, the smooth muscle constructs contracted and relaxed, which is similar to the actual behavior of airway smooth muscle. Nesmith et al. developed a human airway musculature on a chip using PDMS chip body and normal human bronchial smooth muscle [122]. To recapitulate the asthmatic airway caused by the immune response on this chip, interleukin 13 (IL-13) was treated, which is predominant in airway patients, and hypercontraction in response was observed in the smooth muscle lamellae.

As shown above, various studies have been conducted to emulate the in vivo relevant architectures and functions of the airway, and different types of airway models have been developed, from simple structure of spheroid to complex structures of transwell or chip platform. However, in the case of simple structure, it is easy to produce, but it has limitations in recapitulating the actual environments of airway. On the other hand, airway-on-a-chip enables recapitulation of diverse elements of native airway such as air–liquid interface or microfluids, but there have been difficulties in generating each device, causing poor reproducibility. Furthermore, even though the airway ECM is directly linked to its function, this is rarely considered, with type I collagen being used in almost all studies for the development of the airway model. Although studies using decellularized airway have been reported, these cases are not suitable for applying to in vitro models targeting airway mucosal tissues, mainly because the materials are prepared with preserving even parts of cartilage for the purpose of transplantation. Now, as a recent way to address the situation, we will discuss 3D bioprinting technology and tracheal mucosa dECM (tmdECM) bioink to create a sophisticated airway model [113].

3D printing technique can overcome the abovementioned limitations because it is easy to make microchannels where air or microfluids can flow and to position various cell compositions in the desired location. In addition, the use of decellularized airway mucosal tissue-derived bioink has the benefits of reproducing the complex ECM environment of the airway. In this regard, Park et al. developed functional airway-on-a-chip using 3D bioprinting technology and tmdECM bioink (Fig. 7.26) [113].

First, bioink for direct bioprinting was prepared by decellularization of airway mucosal tissue. To confirm the functionality of tmdECM bioink, tube formation assay was performed first using human dermal microvascular endothelial cells (hDMECs), and tube formation efficiency was very high when tmdECM was used alone, compared to when mixed with Matrigel (Fig. 7.27a). The reason for this is due to the inherent VEGF of tmdECM bioink. In addition, the expression of vascular markers such as PECAM and VE-cadherin was significantly increased when tmdECM was used, and it was proved that tmdECM was more optimized for developing the vascular part for airway in vitro model compared to other materials, such as other tissue-derived dECM or type I collagen. To prepare 3D vascular platform,

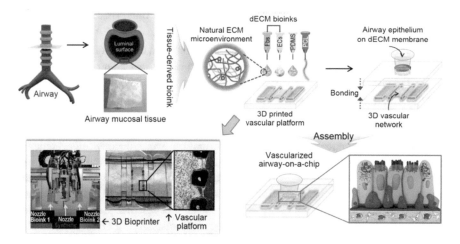

Fig. 7.26 Schematic of 3D bioprinting of airway-on-a-chip using tracheal mucosa dECM bioink. (Reproduced with permission from [113])

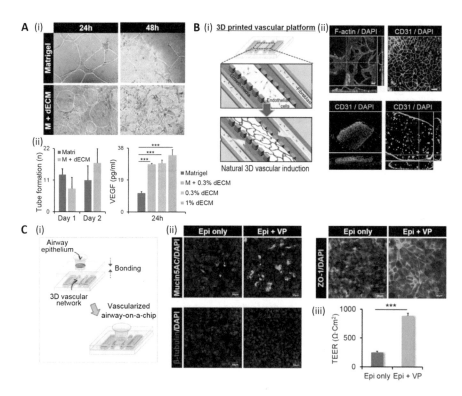

Fig. 7.27 Functionality of tracheal mucosa dECM (tmdECM) bioink and airway-on-a-chip with 3D vascular platform. (Reproduced with permission from [113]). (**a**) Confirmation of high vasculogenic performance of tmdECM bioink by (i) morphology, (ii) tube formation, and (iii) the amount of inherent vascular endothelial growth factors (VEGFs). (**b**) (i) Schematic of 3D-printed vascular platform and (ii) observation of hollow-shaped or interconnected vascular network by staining CD31. (**c**) (i) Assembly of airway epithelium part and 3D vascular platform. (ii) Validation of matured behaviors of epithelium on the 3D vascular platform by higher expression of mucin 5AC, β-tubulin, and ZO1 and (iii) increased transepithelial electrical resistance (TEER) value

the width and height of the reservoir for each channel and cell culture medium were designed to optimize the interaction and function of hDMECs and lung fibroblast. Then, the chip body of a 3D vascular platform was printed using PCL and PDMS via one-step process, and both cells' encapsulated tmdECM bioinks were extruded precisely into the predetermined respective locations (Fig. 7.27b (i)). In the 3D vascular platform, it was confirmed that the expression of CD31 (PECAM), functional vascular network formation, and tube formation were successfully achieved (Fig. 7.27 (ii)). Airway epithelium was prepared by ALI culture on the membrane of tmdECM, and it was assembled with 3D vascular platform to be 3D vascularized airway-on-a-chip (Fig. 7.27c (i)). The expression of mucin 5AC, β-tubulin, and ZO-1 in the airway epithelial part was higher when 3D vascular platform supported it (Fig. 7.27c (ii)) and transepithelial electrical resistance (TEER) value also increased in the same group (Fig. 7.27c (iii)).

As such, when the airway model is fabricated using 3D bioprinting technology and tmdECM bioink, a sophisticated airway microstructure, location of various cells, and biochemical microenvironment can be recapitulated in one device. Moreover, since the same airway model can be produced repeatedly with the same quality if only some printing parameters are set in the initial phase of fabrication, it can be applied very efficiently in terms of the drug development process.

Many attempts have been made to develop a sophisticated airway model to study physiology and pathology of the airway. We have covered from simple spheroid 3D culture to complicated 3D-printed airway-on-a-chip, and there were also studies to fabricate airway smooth muscle as a flexible and tunable layer. Moreover, it has also been shown that 3D environment, ALI, microfluidic stimuli, and cellular compositions are significant for reproducing airway functions. However, taking into consideration the actual airway, where the mucosa and submucosa, which seem to be separated from each other, interact very closely, many efforts are still urgently required to reproduce the complicated pathophysiological environments of the airway.

7.7   Cornea

7.7.1   Structural and Physiological Features of Cornea Tissue

The human eye is a highly complex organ with widely exposed white sclera surrounding a darker-colored iris (Fig. 7.28) [123]. A high ratio of exposed sclera in human allows for greater eye movement for larger field of view. Although an eye is composed of numerous important components, this chapter will more focus on the cornea of the eye providing high transparency for light transmission and refraction. The human cornea (~500 μm in thickness with a horizontal diameter of ~12 mm) is comprised of five distinct transparent layers: the outermost layer of corneal epithelium (~10% thickness), Bowman's layer (~1.5% thickness), stromal (~85% thickness), Descemet's membrane (~1.5% thickness), and lastly a single layer of

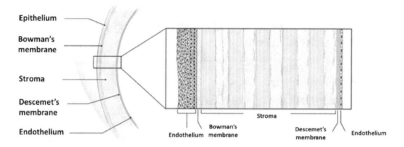

Fig. 7.28 Schematic drawing of five distinct layers in the human cornea. (Reproduced with permission from [123])

endothelial cells (~2% thickness) that lines that posterior cornea. In particular, its high transparency is a result of constant refractive indices in all of its constituent cells.

The surface of corneal epithelium is covered by an overlaying thin layer of tear film made up of Meibomian lipids, which provides the first defender to the external environment. The corneal epithelial cells have an average life span of 7–10 days and undergo routine proliferation, apoptosis, and desquamation. The basal cells are attached on an underlying basement membrane (0.05 μm thickness) comprised of type VI collagen, laminin, and fibronectin secreted by the basal cells and possess lateral intercellular junctions to facilitate strong attachment between the epithelium and underlying corneal layers.

The corneal stroma accounts for approximately 80–85% of the corneal thickness and is predominantly made up of aligned type I collagen fibrils that are embedded in a proteoglycan-rich hydrated matrix [124]. The corneal collagens are organized into uniformly spaced fibrils that form large fiber bundles that are typically 0.2 mm broad and 2 μm thick. Furthermore, the presence of crimps within the collagen fibrils in combination with the elastic fibers makes the cornea more pliable and absorbs intraocular pressure fluctuation. These unique features within the corneal stroma offer structural support to maintain the shape of the anterior corneal surface. The thin corneal collagen fibrils serve as the essential load-bearing constituents of the lamellae and confer high transparency to the cornea. The flattened corneal keratinocytes reside in the stroma region at an average density of 2.0–2.4 million cells/cm^3; an interconnected keratinocyte network is formed via direct contact of dendrites between neighboring cells with a higher cell density in the upper stroma region. Upon injury, the keratinocytes residing in the stroma region are activated to synthesize type I, V, and VI collagen and keratin sulfate.

The corneal endothelium is a thin, and the innermost, layer of cornea; the endothelial cells flatten over time to form a monolayer, and the basal surface of the endothelium contains numerous hemidesmosomes [125]. The primary function of the endothelium layer is to maintain the stroma (78% water content) by pumping excess fluid out of the stroma. This pump function is highly dependent on the endothelial cell morphology. The number of endothelial cells decreases with age; the remaining cells have the capacity to elongate and occupy the void left behind by the degenerated endothelial cells.

7.7.2   3D Bioprinting of the Cornea

The most challenging obstacle in tissue-engineered cornea lies on the fabrication of the corneal stroma with unique lamella arrangement [126]; complex structures such as the interweaving fibrils at both the anterior and posterior stroma and the high uniformity of the nanoscale fibrils are arranged in orthogonal direction. 3D bioprinting is an attractive biofabrication platform, enabling fabrication with the high resolution required for the tissue engineering of corneal stroma with varying cornea curvature of high transparency [127]. Some researchers have employed two-photon polymerization printing (2PP) and multiphoton excitation to align collagen fibrils in the nanoscale or to potentially replicate curvature of corneal structures [128]. However, laser-based printing techniques have been seldom applied for the field of corneal tissue engineering due to the difficulty of their use. More recently, an extrusion-based bioprinting technique was applied for 3D bioprinting a corneal stroma equivalent that resembled the structure of the native human by using a suitable supporting structure corneal stroma [126]. The results showed that the fabricated curved cornea-mimicking structure included high cell viability both at day 1 post-printing (>90%) and at day 7 (83%) (Fig. 7.29).

Despite this technical advance, these outcomes were not fully supporting corneal features including high transparency and tissue-specific microenvironment. To better mimic these biophysical microenvironment, some studies have already introduced thin collagen fibrils and glycosaminoglycan fibrils derived from the decellularized corneal tissues [129]; however, their studies were somewhat limited to the material itself rather than a structural integration of fibril alignments by using a certain technique. As an alternative to overcome this issue, our group developed a decellularized corneal bioink to be used together with extrusion-based 3D bioprinting technique [130]. More importantly, by numerically calculating the wall shear forces exerted on the shear-thinning bioink upon 3D printing, the degree of collagen fibrils was investigated (Fig. 7.30a). In particular, the collagen fibrils in the fabricated structures were aligned under 25 gage of nozzle while maintaining keratocytes' features, leading to high transparency. After 4 weeks in vivo, the collagen fibrils remodeled upon the printing path created a lattice pattern that is similar to the structure of native human cornea as shown in Fig. 7.30b. Collectively, it is obvious that 3D bioprinting of cornea with dECM bioink is promising for better recapitulation of complex architectures of native human cornea. However, current studies had merely focused on cornea stromal layers. Hence, further studies will have to incorporate other layers such as endothelium and epithelium for better structural and physiological recapitulation.

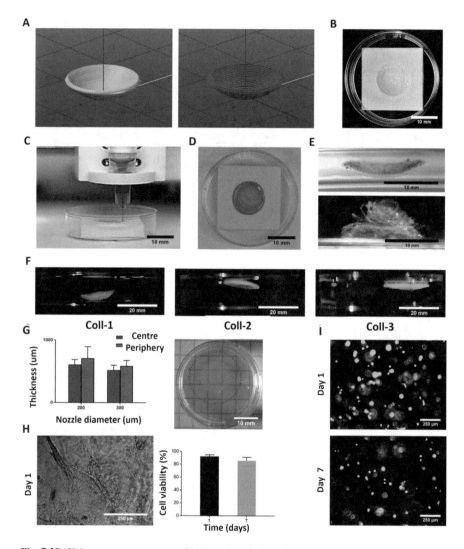

Fig. 7.29 Using support structure to facilitate the printing of a corneal structure with 3% alginate (nozzle diameter = 200 μm) and optimization of bioinks for corneal 3D bioprinting. (**a**) Digital cornea is imported to the computer driving the 3D printer software slices, and a preview of the concentric directionality of print is displayed. (**b**) The support structure is coated with freeform reversible embedding of suspended hydrogel (FRESH) to facilitate the 3D bioprinting of corneal structures. (**c**) View of the 3D bioprinting process. (**d**) Image of 3D-bioprinted corneal structure captured prior to incubation. (**e**) FRESH is aspirated after 8 min of incubation, and corneal structure is carefully removed from support, but begins to unravel 1 day post-printing once keratocytes were combined with the alginate bioink. (**f**) Images of corneal structures 3D bioprinted from composite bioinks. (**g**) Relationship between nozzle diameter and printed thickness of corneal structures (left) and depiction of transparency of corneal structure 3D bioprinted from Coll-1 bioink (right). (**h**) Bright-field image of 3D-bioprinted corneal structure containing cells at day 1 (left) and cell viability measurements over 7 days (right). (**i**) Representative live/dead stain images using fluorescence microscopy at days 1 and 7 after 3D bioprinting in Coll-1. (Reproduced with permission from [126])

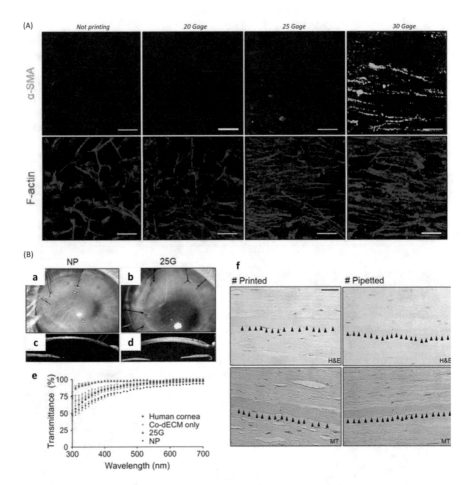

Fig. 7.30 (**a**) Cellular behavior of the differentiated keratocytes encapsulated in bioink along with the shear stresses; 25 gage showed the highest alignment of collagen fibrils without a-SMA expression (scale bar: 50 μm). (**b**) In vivo analysis showing cornea-specific features; optical images from slit lamp examination (a, b); 2D cross-sectional OCT images (c, d); visible light transmittance spectra (e); histological images of transplanted samples (f). Scale bar: 100 μm. (Reproduced with permission from [130])

7.8   Skeletal Muscle

7.8.1   Anatomy and Function of Skeletal Muscle

Skeletal muscles represent approximately 45% of body weight, and most of them are attached to the bones by bundles of collagen fibers known as tendons [131]. Unlike the other two types of muscle tissues in the human body (cardiac muscle and smooth muscle), skeletal muscle can voluntarily relax and contract under the

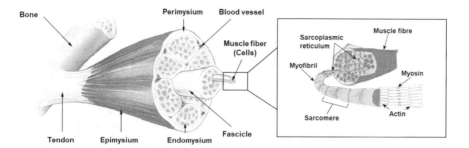

Fig. 7.31 Schematic diagram of skeletal muscle anatomy [131]. (Copyright © 2001 Benjamin Cummings, an imprint of Addison Wesley Longman)

control of the somatic nerve system [132]. Over 600 different skeletal muscles are involved in skeletal support, stability, locomotion, and dynamic events, including regulation of metabolism [133]. For the receipt of nutrients and oxygen as well as the removal of wastes, skeletal muscles are also connected to vascular networks.

Skeletal muscles are composed of multiple bundles (fascicles) of cells bonding together (i.e., muscle fibers). Muscle fibers are formed by the fusion of multiple developmental myoblasts in a process known as myogenesis [134]. The fibers and muscles are wrapped by thin layers of connective tissues called endomysium, which is axially organized and gathered in a bundle covered by another protective connective tissue named perimysium. Multiple bundles are arranged together and form the muscle, covered by the epimysium, another connective tissue (Fig. 7.31).

Under the microscope, the skeletal muscle fibers appear striated due to the alignment of repeated basic units called sarcomeres, which are composed of myosin and actin proteins. These units are important for the contractile function of skeletal muscles. In addition to sarcomeres, the other two critical regulatory proteins, troponin and tropomyosin, are necessary for muscle contraction [135]. Once the muscle cells are sufficiently stimulated, the sarcoplasmic reticulum releases calcium ions, which interact with troponin. Calcium-bound troponin undergoes a conformational change that causes the movement of tropomyosin, subsequently exposing the myosin-binding sites on actin. This process allows for myosin and actin ATP-dependent cross-bridge cycling and for shortening of the muscles.

Healthy skeletal muscles possess the regenerative ability to rehabilitate themselves from small lacerations and wear caused by everyday activity owing to the presence of resident muscle stem cells called satellite cells. However, such a capacity becomes attenuated, due to senescence, genetic diseases, or volumetric muscle loss (VML), which can result in debilitating myopathies that drastically impact the quality of life or shorten life span [136–138]. Therefore, tissue-engineered human skeletal muscle has emerged for regenerative implants for VML treatments and as an in vitro platform for knowledge accumulation of muscle biology and prediction of drug effects.

7.8.2   3D Bioprinting of In Vitro Skeletal Muscle Model

Engineering muscle fibers in vitro requires the culture of myoblasts in an anisotropic environment for supporting cell alignment, promoting cell fusion, and accelerating myogenesis. A plethora of methods have been developed to organize cells, including culturing cells on grooves/ridges patterned substrate or nanofibers [139–142], applying anchors to control axial hydrogel compaction [143], and mechanical/electrical/magnetic stimulations [144]. However, these approaches are limited in terms of inducing precise 3D spatial cell alignment. In this regard, the 3D bioprinting technique holds an extraordinary advantage because of its capacity for providing high precision in cell and matrix positioning for rapidly fabricating desired organized structures.

A number of pioneering studies have focused on the 3D fabrication of axially aligned guiding microfibers that can lead to the formation of muscle fascicles from the post-seeded myoblasts. In one example, microfibrous polycaprolactone (PCL) bundles were fabricated using a melt-printing system to print a poly(vinyl alcohol) (PVA)/PCL (ratio 3:7) solution at 85 °C with a 350 µm nozzle [145]. After the removal of the sacrificial PVA in water, the PCL structure was coated with 0.5% collagen, which was crosslinked by 1-ethyl-3-(3-dimethylaminopropyl)carbodiimide (EDC) for 30 min, followed by freeze-drying for 12 h (Fig. 7.32a). The leaching of microfibrillated PVA resulted in a PCL scaffold with an aligned microfibrous patterned surface. The seeded myoblast (C2C12) cells are organized following such a longitudinal pattern and infiltrated between the microfibers, leading to the formation of a living scaffold mimicking the muscle bundle section. The myosin heavy chain (MHC), a motor protein of muscle thick filaments that regulates muscle functions, was found to be well-developed.

In addition to seeding cells on the micropatterned scaffold, bioinks have been employed to directly print cells for promoting cell survival and activity. Using this method, a research group has attempted to engineer a muscle-tendon unit. Using a composite bioink made of gelatin, fibrinogen, and hyaluronic acid, they bioprinted C2C12 on a previously printed polyurethane (PU) aligned fibrous scaffold that has an elasticity similar to muscle tissue [147]. Similarly, NIH3T3 cells were printed on a PCL aligned fibrous scaffold that mimics the stiffness of tendon. These two scaffolds were overlapped (10%) to imitate the interface between the muscle and tendon. High cell viability (over 94%) was observed after printing. In addition, upon 7 days of culture, C2C12 cells expressed desmin and MHC, while the fibroblasts secreted type I collagen, which is a major ECM of tendon tissue. In a further step, this group 3D bioprinted a human-scale skeletal muscle tissue using cell-laden hydrogel supported by previously printed PCL pillars [146]. The results indicated cell alignments at day 3 and myotube formation after 7 days cultured in differentiation medium (Fig. 7.32b). Except for this composite bioink, other biomaterials such as fibrin [148, 149], silk fibroin [150], type I collagen [151], and Matrigel [152] have been applied to engineer the muscle tissues in vitro. However, these materials are limited in terms either of poor printability or of confined capacity of guiding

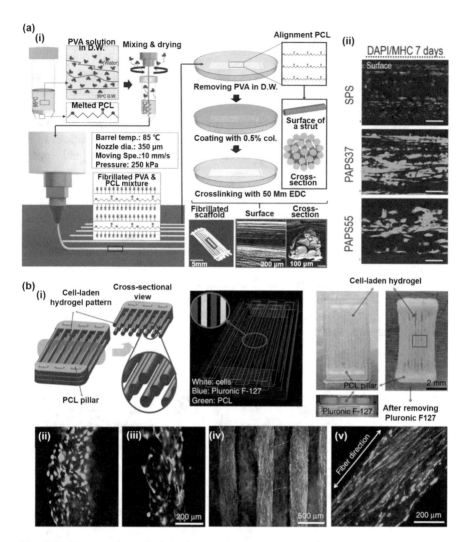

Fig. 7.32 Tissue-engineered skeletal muscle tissue using 3D bioprinting technique. (**a**) (i) Schematic images of fabrication process and optical/SEM images of the hybrid microfibrillated PCL/collagen scaffold. (ii) Immunofluorescence staining images of myosin heavy chain (MHC) (blue = nuclei; green = MHC) after 7 days of cell culture on the scaffold with different designs. (Reproduced with permission from [145]). (**b**) (i) 3D-bioprinted human-scale skeletal muscle tissue constructs by printing cell-laden hydrogel fibers that are supported by PCL pillars. (ii) Without PCL pillar and (iii) with PCL pillar, the encapsulated cells showed different levels of longitudinal organizations along the printing direction. (iv) The live/dead staining of the encapsulated cells in the fiber structure indicates high cell viability after the printing process (green, live cells; red, dead cells). (v) Immunofluorescent staining for MHC of the 3D-printed muscle organization after 7-day differentiation. (Reproduced with permission from [146])

cellular functions of myoblasts. Therefore, the challenge of 3D bioprinting an in vitro muscle model that can exhibit intrinsic morphology and function of native tissue remains.

The ECM of natural tissue provides cells with microenvironmental niche and complex cues that modulate critical cellular processes, including dividing, migration, and differentiation during tissue growth and development. Therefore, an optimal bioink needs to replicate the original composition of native ECMs. However, it is difficult to reproduce such complexity by designing a compound of various ECM constituent due to the limited understanding of ECM specifications and limited manufacturing techniques.

To overcome this challenge, a decellularized skeletal muscle extracellular matrix (mdECM)-based bioink has been developed to reflect the natural composition of muscular ECM [153]. The mdECM was derived from porcine tibialis anterior muscle based on a combination of physical (rinse), chemical (sodium dodecyl sulfate and isopropanol), and enzymatic (DNAse) treatments. After acidic digestion, the resulting muscular tissue-specific bioink with a concentration of 1% showed suitable rheological properties for bioprinting, including shear-thinning behavior and thermal sol–gel transition. Meanwhile, the appropriate modulus of the bioink enables the direct 3D bioprinting of filaments with different patterns (e.g., parallel lines with varying diameters, diamonds, and chains) at 18 °C (Fig. 7.33a). After

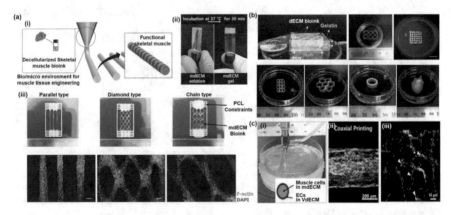

Fig. 7.33 3D bioprinting of skeletal muscle tissue using tissue-specific bioink. (**a**) (i) Schematic diagram of 3D bioprinting of mdECM bioink for the construction of muscle tissue. (ii) Thermal sol–gel transition of 1% mdECM bioink. (iii) The bioink showed suitable bioink for plotting complex patterned cell-laden constructs. (Reproduced with permission from [153]). (**b**) 3D cell-printed thick- and large-volume construct in the granule-based reservoir. This method allows for the fabrication of a variety of complex stacked structures. (**c**) (i) 3D bioprinting of prevascularized muscle constructs through a coaxial nozzle with the assistance of reservoir. (ii) The spatially compartmentalized construct permits the differentiation of each cell type (red, MHC; green, CD31) encapsulated in relevant tissue-specific bioinks. (iii). The staining image shows the innervation in the coaxially bioprinted muscle tissue (green: TUJ1, a neuron cell marker). (Reproduced with permission from [154])

crosslinking at 37 °C and culture for 7 days, the 3D-bioprinted muscle constructs showed significantly higher cell viability (C2C12) and improved cell proliferation compared to the identical structures fabricated using type I collagen bioink. Using a 3D-printed anchor, the encapsulated cells axially aligned along the filaments. Upon the induced cell differentiation, increased myogenic gene expression was detected in the mdECM group compared to that in the collagen group. This superiority of promoting myoblasts' functionalities could be attributed to the biomimetic ECM microenvironments inherited from native muscle tissue. The biochemical analysis demonstrated that the mdECM not only preserves major muscular ECM components such as laminin, collagen, and glycosaminoglycans but also includes agrin that allowed for the pre-patterning of acetylcholine receptors.

Despite the successful exploration of muscular tissue-specific bioink, this material is, similar to other hydrogels, limited in the poor printability and shape fidelity for stacking 3D tall constructs. Therefore, it is difficult to use it for 3D bioprinting the freestanding volumetric muscle tissues with physiologically relevant geometries. To overcome this hurdle, a further study developed a reservoir-bath-assisted printing method [154]. This reservoir is prepared using small-sized gelatin granules (181.21 ± 73.38 μm) and poly(vinyl alcohol) (PVA). While the presence of gelatin granule renders Bingham plastic behavior of the reservoir for maintaining the position of the printed bioink, the addition of PVA plays the role of a coagent to rapidly polymerize the mdECM bioink. Using this configuration, it is possible to 3D bioprint various designs of volumetric tissue constructs, even with multiple cell types (Fig. 7.33b). Therefore, this novel 3D bioprinting strategy might be applicable to fabricating other tissue analogs with complex architectures, such as the heart, kidney, and brain.

With the tissue-specific bioink and versatile 3D bioprinting strategy, the volumetric muscular tissue was successfully constructed. To facilitate the supply of oxygen and nutrients, the constructed ($15 \times 6 \times 4$ mm, length $\times$ width $\times$ height) tissue was designed as multiple muscle constructs with microchannels in between (200 μm). After 7 days of culture and differentiation induction, the muscle construct showed patterned myoblasts and formation of myotubes. Moreover, to improve the functional recovery, a prevascularized muscle construct that imitates the hierarchical architecture of vascularized muscle was fabricated by coaxial printing of both cell-laden mdECM (C2C12) and vascular decellularized extracellular matrix (VdECM) (HUVEC) bioink (Fig. 7.33c). This spatial printing of tissue-specific bioinks offers organized microenvironmental cues to differentiate each cell type, which effectively improves vascularization and muscular functions. This 3D-bioprinted functional volumetric skeletal muscle tissue provides a blueprint for the future development of human-scaled muscle tissue for clinical VML injury treatment, drug effect prediction, and disease pathogenesis study.

References

1. Kanitakis J. Anatomy, histology and immunohistochemistry of normal human skin. Eur J Dermatol. 2002;12(4):390–9. quiz 400–1.
2. Ng WL, Wang S, Yeong WY, Naing MW. Skin bioprinting: impending reality or fantasy? Trends Biotechnol. 2016;34(9):689–99.
3. Hudson TJ. Skin barrier function and allergic risk. Nat Genet. 2006;38(4):399.
4. Driskell RR, Watt FM. Understanding fibroblast heterogeneity in the skin. Trends Cell Biol. 2015;25(2):92–9.
5. Mathes SH, Ruffner H, Graf-Hausner U. The use of skin models in drug development. Adv Drug Deliv Rev. 2014;69:81–102.
6. Lee W, Debasitis JC, Lee VK, Lee J-H, Fischer K, Edminster K, Park J-K, Yoo S-S. Multi-layered culture of human skin fibroblasts and keratinocytes through three-dimensional free-form fabrication. Biomaterials. 2009;30(8):1587–95.
7. Lee V, Singh G, Trasatti JP, Bjornsson C, Xu X, Tran TN, Yoo S-S, Dai G, Karande P. Design and fabrication of human skin by three-dimensional bioprinting. Tissue Eng Part C Methods. 2013;20(6):473–84.
8. Koch L, Deiwick A, Schlie S, Michael S, Gruene M, Coger V, Zychlinski D, Schambach A, Reimers K, Vogt PM. Skin tissue generation by laser cell printing. Biotechnol Bioeng. 2012;109(7):1855–63.
9. Koch L, Kuhn S, Sorg H, Gruene M, Schlie S, Gaebel R, Polchow B, Reimers K, Stoelting S, Ma N. Laser printing of skin cells and human stem cells. Tissue Eng Part C Methods. 2009;16(5):847–54.
10. Pourchet LJ, Thepot A, Albouy M, Courtial EJ, Boher A, Blum LJ, Marquette CA. Human skin 3D bioprinting using scaffold-free approach. Adv Healthc Mater. 2017;6(4):1601101.
11. Kim BS, Kim H, Gao G, Jang J, Cho D-W. Decellularized extracellular matrix: a step towards the next generation source for bioink manufacturing. Biofabrication. 2017;9(3):034104.
12. Kim BS, Kwon YW, Kong J-S, Park GT, Gao G, Han W, Kim M-B, Lee H, Kim JH, Cho D-W. 3D cell printing of in vitro stabilized skin model and in vivo pre-vascularized skin patch using tissue-specific extracellular matrix bioink: a step towards advanced skin tissue engineering. Biomaterials. 2018;168:38–53.
13. Kim BS, Gao G, Kim JY, Cho DW. 3D cell printing of perfusable vascularized human skin equivalent composed of epidermis, dermis, and hypodermis for better structural recapitulation of native skin. Adv Healthc Mater. 2019;8(7):1801019.
14. Kim BS, Lee J-S, Gao G, Cho D-W. Direct 3D cell-printing of human skin with functional transwell system. Biofabrication. 2017;9(2):025034.
15. Pappano AJ, Wier WG. Cardiovascular physiology-E-book: Mosby physiology monograph series. Amsterdam: Elsevier Health Sciences; 2018.
16. Michiels C. Endothelial cell functions. J Cell Physiol. 2003;196(3):430–43.
17. Owens GK. Regulation of differentiation of vascular smooth muscle cells. Physiol Rev. 1995;75(3):487–517.
18. Witter K, Tonar Z, Schöpper H. How many layers has the adventitia?–structure of the arterial tunica externa revisited. Anat Histol Embryol. 2017;46(2):110–20.
19. Wu W, DeConinck A, Lewis JA. Omnidirectional printing of 3D microvascular networks. Adv Mater. 2011;23(24):H178–83.
20. Kolesky DB, Homan KA, Skylar-Scott MA, Lewis JA. Three-dimensional bioprinting of thick vascularized tissues. Proc Natl Acad Sci. 2016;113(12):3179–84.
21. Homan KA, Kolesky DB, Skylar-Scott MA, Herrmann J, Obuobi H, Moisan A, Lewis JA. Bioprinting of 3D convoluted renal proximal tubules on perfusable chips. Sci Rep. 2016;6:34845.
22. Lee VK, Kim DY, Ngo H, Lee Y, Seo L, Yoo S-S, Vincent PA, Dai G. Creating perfused functional vascular channels using 3D bio-printing technology. Biomaterials. 2014;35(28):8092–102.

23. Bertassoni LE, Cecconi M, Manoharan V, Nikkhah M, Hjortnaes J, Cristino AL, Barabaschi G, Demarchi D, Dokmeci MR, Yang Y. Hydrogel bioprinted microchannel networks for vascularization of tissue engineering constructs. Lab Chip. 2014;14(13):2202–11.

24. Miller JS, Stevens KR, Yang MT, Baker BM, Nguyen D-HT, Cohen DM, Toro E, Chen AA, Galie PA, Yu X. Rapid casting of patterned vascular networks for perfusable engineered three-dimensional tissues. Nat Mater. 2012;11(9):768.

25. Hinton TJ, Jallerat Q, Palchesko RN, Park JH, Grodzicki MS, Shue H-J, Ramadan MH, Hudson AR, Feinberg AW. Three-dimensional printing of complex biological structures by freeform reversible embedding of suspended hydrogels. Sci Adv. 2015;1(9):e1500758.

26. Bhattacharjee T, Zehnder SM, Rowe KG, Jain S, Nixon RM, Sawyer WG, Angelini TE. Writing in the granular gel medium. Sci Adv. 2015;1(8):e1500655.

27. Hinton TJ, Hudson A, Pusch K, Lee A, Feinberg AW. 3D printing PDMS elastomer in a hydrophilic support bath via freeform reversible embedding. ACS Biomater Sci Eng. 2016;2(10):1781–6.

28. Song KH, Highley CB, Rouff A, Burdick JA. Complex 3D-printed microchannels within cell-degradable hydrogels. Adv Funct Mater. 2018;28:1801331.

29. Norotte C, Marga FS, Niklason LE, Forgacs G. Scaffold-free vascular tissue engineering using bioprinting. Biomaterials. 2009;30(30):5910–7.

30. Itoh M, Nakayama K, Noguchi R, Kamohara K, Furukawa K, Uchihashi K, Toda S, Oyama J-i, Node K, Morita S. Scaffold-free tubular tissues created by a bio-3D printer undergo remodeling and endothelialization when implanted in rat aortae. PLoS One. 2015;10(9):e0136681.

31. Gao Q, He Y, Fu J-z, Liu A, Ma L. Coaxial nozzle-assisted 3D bioprinting with built-in microchannels for nutrients delivery. Biomaterials. 2015;61:203–15.

32. Zhang Y, Yu Y, Akkouch A, Dababneh A, Dolati F, Ozbolat IT. In vitro study of directly bioprinted perfusable vasculature conduits. Biomater Sci. 2015;3(1):134–43.

33. Jia W, Gungor-Ozkerim PS, Zhang YS, Yue K, Zhu K, Liu W, Pi Q, Byambaa B, Dokmeci MR, Shin SR. Direct 3D bioprinting of perfusable vascular constructs using a blend bioink. Biomaterials. 2016;106:58–68.

34. Liu W, Zhong Z, Hu N, Zhou Y, Maggio L, Miri AK, Fragasso A, Jin X, Khademhosseini A, Zhang YS. Coaxial extrusion bioprinting of 3D microfibrous constructs with cell-favorable gelatin methacryloyl microenvironments. Biofabrication. 2018;10(2):024102.

35. Gao G, Lee JH, Jang J, Lee DH, Kong JS, Kim BS, Choi YJ, Jang WB, Hong YJ, Kwon SM. Tissue engineered bio-blood-vessels constructed using a tissue-specific bioink and 3D coaxial cell printing technique: a novel therapy for ischemic disease. Adv Funct Mater. 2017;27(33):1700798.

36. Gao G, Park JY, Kim BS, Jang J, Cho DW. Coaxial cell printing of freestanding, perfusable, and functional in vitro vascular models for recapitulation of native vascular endothelium pathophysiology. Adv Healthc Mater. 2018;7(23):1801102.

37. Abdel-Misih SR, Bloomston M. Liver anatomy. Surg Clin. 2010;90(4):643–53.

38. van Grunsven LA. 3D in vitro models of liver fibrosis. Adv Drug Deliv Rev. 2017;121:133–46.

39. Stenvall A, Larsson E, Strand S-E, Jönsson B-A. A small-scale anatomical dosimetry model of the liver. Phys Med Biol. 2014;59(13):3353.

40. Fomin ME, Zhou Y, Beyer AI, Publicover J, Baron JL, Muench MO. Production of factor VIII by human liver sinusoidal endothelial cells transplanted in immunodeficient uPA mice. PLoS One. 2013;8(10):e77255.

41. Yin C, Evason KJ, Asahina K, Stainier DY. Hepatic stellate cells in liver development, regeneration, and cancer. J Clin Invest. 2013;123(5):1902–10.

42. Bilzer M, Roggel F, Gerbes AL. Role of Kupffer cells in host defense and liver disease. Liver Int. 2006;26(10):1175–86.

43. Zhang R-R, Zheng Y-W, Li B, Nie Y-Z, Ueno Y, Tsuchida T, Taniguchi H. Hepatic stem cells with self-renewal and liver repopulation potential are harbored in CDCP1-positive subpopulations of human fetal liver cells. Stem Cell Res Ther. 2018;9(1):29.

44. Shu XZ, Ahmad S, Liu Y, Prestwich GD. Synthesis and evaluation of injectable, in situ crosslinkable synthetic extracellular matrices for tissue engineering. J Biomed Mater Res A. 2006;79(4):902–12.
45. Deegan D. Effects of liver extracellular matrix gel stiffness on primary hepatocyte function. PhD Thesis. Wake Forest University. 2015.
46. Cho D-W, Lee H, Han W, Choi Y-J. Bioprinting of liver. 3D bioprinting in regenerative engineering: principles and applications. Boca Raton, FL: CRC Press; 2018.
47. Eddershaw PJ, Beresford AP, Bayliss MK. ADME/PK as part of a rational approach to drug discovery. Drug Discov Today. 2000;5(9):409–14.
48. Lewis JH, Ahmed M, Shobassy A, Palese C. Drug-induced liver disease. Curr Opin Gastroenterol. 2006;22(3):223–33.
49. Kola I, Landis J. Can the pharmaceutical industry reduce attrition rates? Nat Rev Drug Discov. 2004;3(8):711.
50. Sharer JE, Shipley LA, Vandenbranden MR, Binkley SN, Wrighton SA. Comparisons of phase I and phase II in vitro hepatic enzyme activities of human, dog, rhesus monkey, and cynomolgus monkey. Drug Metab Dispos. 1995;23(11):1231–41.
51. Kullak-Ublick GA, Andrade RJ, Merz M, End P, Benesic A, Gerbes AL, Aithal GP. Drug-induced liver injury: recent advances in diagnosis and risk assessment. Gut. 2017;66(6):1154–64.
52. Du Y, Han R, Wen F, San San SN, Xia L, Wohland T, Leo HL, Yu H. Synthetic sandwich culture of 3D hepatocyte monolayer. Biomaterials. 2008;29(3):290–301.
53. Eschbach E, Chatterjee SS, Nöldner M, Gottwald E, Dertinger H, Weibezahn KF, Knedlitschek G. Microstructured scaffolds for liver tissue cultures of high cell density: morphological and biochemical characterization of tissue aggregates. J Cell Biochem. 2005;95(2):243–55.
54. Ise H, Takashima S, Nagaoka M, Ferdous A, Akaike T. Analysis of cell viability and differential activity of mouse hepatocytes under 3D and 2D culture in agarose gel. Biotechnol Lett. 1999;21(3):209–13.
55. Wang Y, Su W, Wang L, Jiang L, Liu Y, Hui L, Qin J. Paper supported long-term 3D liver co-culture model for the assessment of hepatotoxic drugs. Toxicol Res. 2018;7(1):13–21.
56. Wei G, Wang J, Lv Q, Liu M, Xu H, Zhang H, Jin L, Yu J, Wang X. Three-dimensional coculture of primary hepatocytes and stellate cells in silk scaffold improves hepatic morphology and functionality in vitro. J Biomed Mater Res A. 2018;106(8):2171–80.
57. Liu Y, Li H, Yan S, Wei J, Li X. Hepatocyte cocultures with endothelial cells and fibroblasts on micropatterned fibrous mats to promote liver-specific functions and capillary formation capabilities. Biomacromolecules. 2014;15(3):1044–54.
58. Yi H-G, Lee H, Cho D-W. 3D printing of organs-on-chips. Bioengineering. 2017;4(1):10.
59. Toh Y-C, Lim TC, Tai D, Xiao G, van Noort D, Yu H. A microfluidic 3D hepatocyte chip for drug toxicity testing. Lab Chip. 2009;9(14):2026–35.
60. Rennert K, Steinborn S, Gröger M, Ungerböck B, Jank A-M, Ehgartner J, Nietzsche S, Dinger J, Kiehntopf M, Funke H. A microfluidically perfused three dimensional human liver model. Biomaterials. 2015;71:119–31.
61. Bhise NS, Manoharan V, Massa S, Tamayol A, Ghaderi M, Miscuglio M, Lang Q, Zhang YS, Shin SR, Calzone G. A liver-on-a-chip platform with bioprinted hepatic spheroids. Biofabrication. 2016;8(1):014101.
62. Lee H, Cho D-W. One-step fabrication of an organ-on-a-chip with spatial heterogeneity using a 3D bioprinting technology. Lab Chip. 2016;16(14):2618–25.
63. Lee H, Han W, Kim H, Ha D-H, Jang J, Kim BS, Cho D-W. Development of liver decellularized extracellular matrix bioink for three-dimensional cell printing-based liver tissue engineering. Biomacromolecules. 2017;18(4):1229–37.
64. Skardal A, Devarasetty M, Kang H-W, Mead I, Bishop C, Shupe T, Lee SJ, Jackson J, Yoo J, Soker S. A hydrogel bioink toolkit for mimicking native tissue biochemical and mechanical properties in bioprinted tissue constructs. Acta Biomater. 2015;25:24–34.
65. Lee H, Chae S, Kim JY, Han W, Kim J, Choi Y-J, Cho D-W. Cell-printed 3D liver-on-a-chip possessing a liver microenvironment and biliary system. Biofabrication. 2019;11:025001.

66. Du B, Yu M, Zheng J. Transport and interactions of nanoparticles in the kidneys. Nat Rev Mater. 2018;3(10):358–74.

67. Lin NYC, Homan KA, Robinson SS, Kolesky DB, Duarte N, Moisan A, Lewis JA. Renal reabsorption in 3D vascularized proximal tubule models. Proc Natl Acad Sci U S A. 2019;116(12):5399–404.

68. Ali M, Pr AK, Yoo JJ, Zahran F, Atala A, Lee SJ. A photo-crosslinkable kidney ECM-derived bioink accelerates renal tissue formation. Adv Healthc Mater. 2019;8(7):e1800992.

69. Kitsara M, Agbulut O, Kontziampasis D, Chen Y, Menasche P. Fibers for hearts: a critical review on electrospinning for cardiac tissue engineering. Acta Biomater. 2017;48:20–40.

70. Whitaker RH. Anatomy of the heart. Medicine. 2014;42(8):406–8.

71. Reis LA, Chiu LL, Feric N, Fu L, Radisic M. Biomaterials in myocardial tissue engineering. J Tissue Eng Regen Med. 2016;10(1):11–28.

72. Heidenreich PA, Trogdon JG, Khavjou OA, Butler J, Dracup K, Ezekowitz MD, Finkelstein EA, Hong Y, Johnston SC, Khera A, Lloyd-Jones DM, Nelson SA, Nichol G, Orenstein D, Wilson PW, Woo YJ, American Heart C, Association Advocacy Coordinating, C. Stroke, R. Council on Cardiovascular, Intervention, C. Council on Clinical, E. Council on, Prevention, A. Council on, Thrombosis, B. Vascular, C. Council on, C. Critical, Perioperative, Resuscitation, N. Council on Cardiovascular, D. Council on the Kidney in Cardiovascular, S. Council on Cardiovascular, Anesthesia, C. Interdisciplinary Council on Quality of, R. Outcomes. Forecasting the future of cardiovascular disease in the United States: a policy statement from the American Heart Association. Circulation. 2011;123(8):933–44.

73. Das S, Jang J. 3D bioprinting and decellularized ECM-based biomaterials for in vitro CV tissue engineering. J 3D Print Med. 2018;2(2):69–87.

74. Page A, Messer S, Large SR. Heart transplantation from donation after circulatory determined death. Ann Cardiothorac Surg. 2018;7(1):75–81.

75. Pajaro OE, Jaroszewski DE, Scott RL, Kalya AV, Tazelaar HD, Arabia FA. Antibody-mediated rejection in heart transplantation: case presentation with a review of current international guidelines. J Transpl. 2011;2011:351950.

76. Rangarajan A, Hong SJ, Gifford A, Weinberg RA. Species- and cell type-specific requirements for cellular transformation. Cancer Cell. 2004;6(2):171–83.

77. Wilke RA, Lin DW, Roden DM, Watkins PB, Flockhart D, Zineh I, Giacomini KM, Krauss RM. Identifying genetic risk factors for serious adverse drug reactions: current progress and challenges. Nat Rev Drug Discov. 2007;6(11):904–16.

78. Mathur A, Ma Z, Loskill P, Jeeawoody S, Healy KE. In vitro cardiac tissue models: current status and future prospects. Adv Drug Deliv Rev. 2016;96:203–13.

79. Pati F, Gantelius J, Svahn HA. 3D bioprinting of tissue/organ models. Angew Chem. 2016;55(15):4650–65.

80. Vunjak-Novakovic G, Tandon N, Godier A, Maidhof R, Marsano A, Martens TP, Radisic M. Challenges in cardiac tissue engineering. Tissue Eng Part B Rev. 2010;16(2):169–87.

81. Murphy SV, Atala A. 3D bioprinting of tissues and organs. Nat Biotechnol. 2014;32(8):773–85.

82. Choi YJ, Yi HG, Kim SW, Cho DW. 3D cell printed tissue analogues: a new platform for theranostics. Theranostics. 2017;7(12):3118–37.

83. Garreta E, Oria R, Tarantino C, Pla-Roca M, Prado P, Fernández-Avilés F, Campistol JM, Samitier J, Montserrat N. Tissue engineering by decellularization and 3D bioprinting. Mater Today. 2017;20(4):166–78.

84. Jang J, Park JY, Gao G, Cho DW. Biomaterials-based 3D cell printing for next-generation therapeutics and diagnostics. Biomaterials. 2018;156:88–106.

85. Ott HC, Matthiesen TS, Goh SK, Black LD, Kren SM, Netoff TI, Taylor DA. Perfusion-decellularized matrix: using nature's platform to engineer a bioartificial heart. Nat Med. 2008;14(2):213–21.

86. Brown BN, Badylak SF. Extracellular matrix as an inductive scaffold for functional tissue reconstruction. Transl Res. 2014;163(4):268–85.

87. Schoen B, Avrahami R, Baruch L, Efraim Y, Goldfracht I, Elul O, Davidov T, Gepstein L, Zussman E, Machluf M. Electrospun extracellular matrix: paving the way to tailor-made natural scaffolds for cardiac tissue regeneration. Adv Funct Mater. 2017;27(34):1700427.

88. Badylak SF, Freytes DO, Gilbert TW. Extracellular matrix as a biological scaffold material: structure and function. Acta Biomater. 2009;5(1):1–13.

89. Pati F, Song TH, Rijal G, Jang J, Kim SW, Cho DW. Ornamenting 3D printed scaffolds with cell-laid extracellular matrix for bone tissue regeneration. Biomaterials. 2015;37:230–41.

90. Wang X, Yan Y, Zhang R. Rapid prototyping as a tool for manufacturing bioartificial livers. Trends Biotechnol. 2007;25(11):505–13.

91. Derby B. Printing and prototyping of tissues and scaffolds. Science. 2012;338(6109):921–6.

92. Lu T, Li Y, Chen T. Techniques for fabrication and construction of three-dimensional scaffolds for tissue engineering. Int J Nanomedicine. 2013;8:337–50.

93. Pati F, Jang J, Ha DH, Won Kim S, Rhie JW, Shim JH, Kim DH, Cho DW. Printing three-dimensional tissue analogues with decellularized extracellular matrix bioink. Nat Commun. 2014;5:3935.

94. Das S, Kim SW, Choi YJ, Lee S, Lee SH, Kong JS, Park HJ, Cho DW, Jang J. Decellularized extracellular matrix bioinks and the external stimuli to enhance cardiac tissue development in vitro. Acta Biomater. 2019;95:188.

95. Jang J, Park HJ, Kim SW, Kim H, Park JY, Na SJ, Kim HJ, Park MN, Choi SH, Park SH, Kim SW, Kwon SM, Kim PJ, Cho DW. 3D printed complex tissue construct using stem cell-laden decellularized extracellular matrix bioinks for cardiac repair. Biomaterials. 2017;112:264–74.

96. Jang J, Kim TG, Kim BS, Kim SW, Kwon SM, Cho DW. Tailoring mechanical properties of decellularized extracellular matrix bioink by vitamin B2-induced photo-crosslinking. Acta Biomater. 2016;33:88–95.

97. Yu C, Ma X, Zhu W, Wang P, Miller KL, Stupin J, Koroleva-Maharajh A, Hairabedian A, Chen S. Scanningless and continuous 3D bioprinting of human tissues with decellularized extracellular matrix. Biomaterials. 2019;194:1–13.

98. Hyde DM, Hamid Q, Irvin CG. Anatomy, pathology, and physiology of the tracheobronchial tree: emphasis on the distal airways. J Allergy Clin Immunol. 2009;124(6):S72–7.

99. Des Jardins TR. Cardiopulmonary anatomy & physiology: essentials for respiratory care. Clifton Park, NY: Delmar Thomson Learning; 2002.

100. Hicks GH. Cardiopulmonary anatomy and physiology. Philadelphia, PA: WB Saunders Company; 2000.

101. Fahy JV, Dickey BF. Airway mucus function and dysfunction. N Engl J Med. 2010;363(23):2233–47.

102. Finkbeiner WE, Zlock LT, Mehdi I, Widdicombe JH. Cultures of human tracheal gland cells of mucous or serous phenotype. In Vitro Cell Dev Biol Anim. 2010;46(5):450–6.

103. Rogers A, Dewar A, Corrin B, Jeffery P. Identification of serous-like cells in the surface epithelium of human bronchioles. Eur Respir J. 1993;6(4):498–504.

104. Knight DA, Holgate ST. The airway epithelium: structural and functional properties in health and disease. Respirology. 2003;8(4):432–46.

105. Paulsson M. Basement membrane proteins: structure, assembly, and cellular interactions. Crit Rev Biochem Mol Biol. 1992;27(1-2):93–127.

106. Fraser RS. Histology and gross anatomy of the respiratory tract. In: Physiologic basis of respiratory disease. Hamilton, ON: BC Decker Inc; 2005. p. 1–14.

107. Shikina T, Hiroi T, Iwatani K, Jang MH, Fukuyama S, Tamura M, Kubo T, Ishikawa H, Kiyono H. IgA class switch occurs in the organized nasopharynx-and gut-associated lymphoid tissue, but not in the diffuse lamina propria of airways and gut. J Immunol. 2004;172(10):6259–64.

108. Widdicombe JH, Wine JJ. Airway gland structure and function. Physiol Rev. 2015;95(4):1241–319.

109. Jan De Backer C, Marchal T, Director HI. Taming the cost of respiratory drug development. ANSYS Adv. 2010;IV:10.

110. Adams CP, Brantner VV. Estimating the cost of new drug development: is it really $802 million? Health Aff. 2006;25(2):420–8.

111. Mestre-Ferrandiz J, Sussex J, Towse A. The R&D cost of a new medicine. Monographs. London: Office of Health Economics; 2012.

112. Cook D, Brown D, Alexander R, March R, Morgan P, Satterthwaite G, Pangalos MN. Lessons learned from the fate of AstraZeneca's drug pipeline: a five-dimensional framework. Nat Rev Drug Discov. 2014;13(6):419.

113. Park JY, Ryu H, Lee B, Ha D-H, Ahn M, Kim S, Kim JY, Jeon NL, Cho D-W. Development of a functional airway-on-a-chip by 3D cell printing. Biofabrication. 2018;11(1):015002.

114. Fessart D, Begueret H, Delom F. Three-dimensional culture model to distinguish normal from malignant human bronchial epithelial cells. Eur Respir J. 2013;42(5):1345–56.

115. Whitcutt MJ, Adler KB, Wu R. A biphasic chamber system for maintaining polarity of differentiation of culture respiratory tract epithelial cells. In Vitro Cell Dev Biol. 1988;24(5):420–8.

116. Myerburg MM, Latoche JD, McKenna EE, Stabile LP, Siegfried JS, Feghali-Bostwick CA, Pilewski JM. Hepatocyte growth factor and other fibroblast secretions modulate the phenotype of human bronchial epithelial cells. Am J Physiol Lung Cell Mol Physiol. 2007;292(6):L1352–60.

117. Blume C, Reale R, Held M, Loxham M, Millar TM, Collins JE, Swindle EJ, Morgan H, Davies DE. Cellular crosstalk between airway epithelial and endothelial cells regulates barrier functions during exposure to double-stranded RNA. Immun Inflamm Dis. 2017;5(1):45–56.

118. Benam KH, Villenave R, Lucchesi C, Varone A, Hubeau C, Lee H-H, Alves SE, Salmon M, Ferrante TC, Weaver JC. Small airway-on-a-chip enables analysis of human lung inflammation and drug responses in vitro. Nat Methods. 2016;13(2):151.

119. Lambert R, Wiggs B, Kuwano K, Hogg J, Pare P. Functional significance of increased airway smooth muscle in asthma and COPD. J Appl Physiol. 1993;74(6):2771–81.

120. Wright D, Sharma P, Ryu M-H, Rissé P-A, Ngo M, Maarsingh H, Koziol-White C, Jha A, Halayko AJ, West AR. Models to study airway smooth muscle contraction in vivo, ex vivo and in vitro: implications in understanding asthma. Pulm Pharmacol Ther. 2013;26(1):24–36.

121. West AR, Zaman N, Cole DJ, Walker MJ, Legant WR, Boudou T, Chen CS, Favreau JT, Gaudette GR, Cowley EA. Development and characterization of a 3D multicell microtissue culture model of airway smooth muscle. Am J Physiol Lung Cell Mol Physiol. 2012;304(1):L4–L16.

122. Nesmith AP, Agarwal A, McCain ML, Parker KK. Human airway musculature on a chip: an in vitro model of allergic asthmatic bronchoconstriction and bronchodilation. Lab Chip. 2014;14(20):3925–36.

123. DelMonte DW, Kim T. Anatomy and physiology of the cornea. J Cataract Refract Surg. 2011;37(3):588–98.

124. Kuwabara T. Current concepts in anatomy and histology of the cornea. Contact Intraocul Lens Med J. 1978;4:101–32.

125. Waring GO III, Bourne WM, Edelhauser HF, Kenyon KR. The corneal endothelium: normal and pathologic structure and function. Ophthalmology. 1982;89(6):531–90.

126. Isaacson A, Swioklo S, Connon CJ. 3D bioprinting of a corneal stroma equivalent. Exp Eye Res. 2018;173:188–93.

127. Ghezzi CE, Rnjak-Kovacina J, Kaplan DL. Corneal tissue engineering: recent advances and future perspectives. Tissue Eng Part B Rev. 2015;21(3):278–87.

128. Ilina O, Bakker G-J, Vasaturo A, Hoffman RM, Friedl P. Two-photon laser-generated microtracks in 3D collagen lattices: principles of MMP-dependent and -independent collective cancer cell invasion. Phys Biol. 2011;8(1):015010.

129. Lynch AP, Ahearne M. Strategies for developing decellularized corneal scaffolds. Exp Eye Res. 2013;108:42–7.

130. Kim H, Jang J, Park J, Lee K-P, Lee S, Lee D-M, Kim KH, Kim HK, Cho D-W. Shear-induced alignment of collagen fibrils using 3D cell printing for corneal stroma tissue engineering. Biofabrication. 2019;11(3):035017.

131. Fox SI. Human physiology. 9th ed. New York, NY: McGraw-Hill Press; 2006.

132. Jolesz F, Sreter FA. Development, innervation, and activity-pattern induced changes in skeletal muscle. Annu Rev Physiol. 1981;43(1):531–52.

133. Ostrovidov S, Salehi S, Costantini M, Suthiwanich K, Ebrahimi M, Sadeghian RB, Fujie T, Shi X, Cannata S, Gargioli C. 3D bioprinting in skeletal muscle tissue engineering. Small. 2019;15:1805530.

134. Kim JH, Jin P, Duan R, Chen EH. Mechanisms of myoblast fusion during muscle development. Curr Opin Genet Dev. 2015;32:162–70.
135. Solaro RJ, Rarick HM. Troponin and tropomyosin: proteins that switch on and tune in the activity of cardiac myofilaments. Circ Res. 1998;83(5):471–80.
136. Alway SE, Myers MJ, Mohamed JS. Regulation of satellite cell function in sarcopenia. Front Aging Neurosci. 2014;6:246.
137. Wallace GQ, McNally EM. Mechanisms of muscle degeneration, regeneration, and repair in the muscular dystrophies. Annu Rev Physiol. 2009;71:37–57.
138. Grogan BF, Hsu JR, Skeletal Trauma Research C. Volumetric muscle loss. J Am Acad Orthopaed Surg. 2011;19:S35–7.
139. Shi X, Ostrovidov S, Zhao Y, Liang X, Kasuya M, Kurihara K, Nakajima K, Bae H, Wu H, Khademhosseini A. Microfluidic spinning of cell-responsive grooved microfibers. Adv Funct Mater. 2015;25(15):2250–9.
140. Gao H, Cao X, Dong H, Fu X, Wang Y. Influence of 3D Microgrooves on C2C12 cell proliferation, migration, alignment, F-actin protein expression and gene expression. J Mater Sci Technol. 2016;32(9):901–8.
141. Ostrovidov S, Shi X, Zhang L, Liang X, Kim SB, Fujie T, Ramalingam M, Chen M, Nakajima K, Al-Hazmi F. Myotube formation on gelatin nanofibers–multi-walled carbon nanotubes hybrid scaffolds. Biomaterials. 2014;35(24):6268–77.
142. Ostrovidov S, Ebrahimi M, Bae H, Nguyen HK, Salehi S, Kim SB, Kumatani A, Matsue T, Shi X, Nakajima K. Gelatin–polyaniline composite nanofibers enhanced excitation–contraction coupling system maturation in myotubes. ACS Appl Mater Interfaces. 2017;9(49):42444–58.
143. Fujie T, Ahadian S, Liu H, Chang H, Ostrovidov S, Wu H, Bae H, Nakajima K, Kaji H, Khademhosseini A. Engineered nanomembranes for directing cellular organization toward flexible biodevices. Nano Lett. 2013;13(7):3185–92.
144. Ostrovidov S, Shi X, Sadeghian RB, Salehi S, Fujie T, Bae H, Ramalingam M, Khademhosseini A. Stem cell differentiation toward the myogenic lineage for muscle tissue regeneration: a focus on muscular dystrophy. Stem Cell Rev Rep. 2015;11(6):866–84.
145. Kim W, Kim M, Kim GH. 3D-printed biomimetic scaffold simulating microfibril muscle structure. Adv Funct Mater. 2018;28(26):1800405.
146. Kang H-W, Lee SJ, Ko IK, Kengla C, Yoo JJ, Atala A. A 3D bioprinting system to produce human-scale tissue constructs with structural integrity. Nat Biotechnol. 2016;34(3):312.
147. Merceron TK, Burt M, Seol Y-J, Kang H-W, Lee SJ, Yoo JJ, Atala A. A 3D bioprinted complex structure for engineering the muscle–tendon unit. Biofabrication. 2015;7(3):035003.
148. Pollot BE, Rathbone CR, Wenke JC, Guda T. Natural polymeric hydrogel evaluation for skeletal muscle tissue engineering. J Biomed Mater Res B Appl Biomater. 2018;106(2):672–9.
149. Matthias N, Hunt SD, Wu J, Lo J, Callahan LAS, Li Y, Huard J, Darabi R. Volumetric muscle loss injury repair using in situ fibrin gel cast seeded with muscle-derived stem cells (MDSCs). Stem Cell Res. 2018;27:65–73.
150. Dixon TA, Cohen E, Cairns DM, Rodriguez M, Mathews J, Jose RR, Kaplan DL. Bioinspired three-dimensional human neuromuscular junction development in suspended hydrogel arrays. Tissue Eng Part C Methods. 2018;24(6):346–59.
151. Cvetkovic C, Raman R, Chan V, Williams BJ, Tolish M, Bajaj P, Sakar MS, Asada HH, Saif MTA, Bashir R. Three-dimensionally printed biological machines powered by skeletal muscle. Proc Natl Acad Sci. 2014;111(28):10125–30.
152. Raman R, Grant L, Seo Y, Cvetkovic C, Gapinske M, Palasz A, Dabbous H, Kong H, Pinera PP, Bashir R. Damage, healing, and remodeling in optogenetic skeletal muscle bioactuators. Adv Healthc Mater. 2017;6(12):1700030.
153. Choi YJ, Kim TG, Jeong J, Yi HG, Park JW, Hwang W, Cho DW. 3D Cell printing of functional skeletal muscle constructs using skeletal muscle-derived bioink. Adv Healthc Mater. 2016;5(20):2636–45.
154. Choi Y-J, Jun Y-J, Kim DY, Yi H-G, Chae S-H, Kang J, Lee J, Gao G, Kong J-S, Jang J. A 3D cell printed muscle construct with tissue-derived bioink for the treatment of volumetric muscle loss. Biomaterials. 2019;206:160.

Chapter 8
Future Outlooks and Conclusions

This textbook has demonstrated that 3D-bioprinted in vitro tissue/organ models using tissue-specific bioinks open up exciting prospects in engineering a more realistic and predictable testing platform. Despite this potential, there are several concerns to be addressed. One of the main concerns is that the productivity and reproducibility of current bioprinted dECM-based tissue models are very limited; only one 3D tissue construct has been built using the 3D bioprinting system. In order to obtain highly accurate and reliable results applicable clinically, high-throughput screening on 3D-bioprinted tissue models should be considered. Because 3D bioprinting tissue models facilitate simultaneous incorporation of a high-contents and high-throughput system, an advanced technique focusing on high-throughput screening would be conducive to accurate prediction.

Clearly, recent advances have enabled researchers to preciously position cells and biomaterials with the purpose of fabrication of functional tissue models. However, most dECM-based in vitro tissue models are focused on a normal state of natural tissues and organs. To better understand physiopathology in an in vivo microenvironment, future research is needed to develop in vitro tissue models that can be used for studying disease mechanisms and personalized drug screening. To this end, patient-specific cell sources, including human iPSC-derived cells and primary diseased cells from patients, might be the main focus.

In addition, further innovations on post-printing culture platform such as bioreactors and microfluidic devices will be needed to assist functional maturation of maintenance. Along with these developments, technology advancement in imaging systems and analyzing tools for accurate observation should be in high demand as well. By applying these dynamic and advanced systems, generation of a human-on-a-chip that several critical tissues are interconnected with each other can be realized to recapitulate a fully integrated platform able to study the interdependent effects of multiple miniaturized organs. Overall, such future efforts will make significant progress of in vitro tissue models using tissue-specific bioinks toward sophisticated precision medicine.

© Springer Nature Switzerland AG 2019
D.-W. Cho et al., *3D Bioprinting*, https://doi.org/10.1007/978-3-030-32222-9_8

As emphasized in this textbook, several works have demonstrated the potential of biofabricated tissue models based on 3D bioprinting. In parallel to this technological advance, dECM biomaterials have been recognized as a well-qualified bioink source that is able to offer realistic microenvironments. A review article even referred to this bioink as "a game changer" toward advanced tissue engineering [1]. Despite the recent advancement in 3D bioprinting techniques, particularly together with dECM bioinks for in vitro tissue biofabrication, a dECM bioink is a mere concoction of various ECM proteins. Thus, dECM bioink is unable to fully replicate the complex ECM, as the highly specific spatial position of each unique protein within the native tissues is disrupted during the bioink formation. Hence, it is essential to improve our understanding of the native ECM by analyzing the distribution and composition of various ECM proteins in the dECM bioinks. An in-depth knowledge of the spatial arrangement of living cells and ECM within tissue constructs along with the development of advanced bioprinting strategies is also necessary to achieve the envisioned and ideal in vitro tissue models. In addition, the use of complementary strategies is crucial to ensure the post-processing from pre-processing of bioprinted tissue constructs into vascularized, functional tissues/organs in vitro. Finally, the potential of bioprinting along with different converging fields of computational, biological, and material sciences makes bioprinting in vitro tissue models a pending reality.

We hope that this textbook can provide a basic step toward comprehensive understanding of bioprinting and in vitro tissue modeling, opening a gate for related beginners considering 3D biofabrication techniques and biomaterials.

Reference

1. Choudhury D, Tun HW, Wang T, Naing MW. Organ-derived decellularized extracellular matrix: a game changer for bioink manufacturing? Trends Biotechnol. 2018;36:787.

Index

© Springer Nature Switzerland AG 2019
D.-W. Cho et al., *3D Bioprinting*, https://doi.org/10.1007/978-3-030-32222-9

Printed in the United States
By Bookmasters